More Praise f

"Goldstein-Rose, a
offers what the world desp
climate change, with the goal of steering the
debate toward solutions that are both technically and
politically practicable."

— Steven Pinker, Professor of Psychology, Harvard University

"Solomon Goldstein-Rose is not just an analyst who brings a grasp of the subject matter to the table, but also a former state elected official who, at a tender age, learned quickly what's needed to make change. His combination of knowledge and experience will reach new audiences and could inject new energy into the debate and give hope to a generation."

—Armond Cohen, Founder and Director, Clean Air Task Force

"A pragmatic, feasible road map for addressing climate change. This is exactly the kind of thinking that is needed now."

—Christina Paxson, President of Brown University

"*The 100% Solution* is an important read for anyone who cares more about addressing climate change than fighting ideological battles."

—Ted Nordhaus, Founder and Director, Breakthrough Institute

"[Solomon's] vision offers a practical path forward and more climate thinkers need to understand his ideas."
—Dan Bosley, former State Representative, architect of the 1997 MA Electric Restructuring Act

"Goldstein-Rose is the rare Millennial author who is passionate about protecting the planet for his and future generations but is also sober and rigorous in his prescriptions to reduce climate-warming emissions . . . *The 100% Solution* is a model for pragmatic and systemic thinking about climate change and proof that activist passion and scholarly rigor can go hand in hand."
—Varun Sivaram, Chief Technology Officer at India's largest renewable energy company, author of *Taming the Sun: Innovations to Harness Solar Energy and Power the Planet*

"With clear text and elegant organization, Goldstein-Rose sets aside apocalyptic visions and maps planetary rescue."
—Professor Steven J. Davis, Associate Professor of Earth System Science, UC–Irvine

"I'm proud to have worked alongside Solomon and seen firsthand what a tremendous force for change he is. He has already proven himself an adept political leader and an effective communicator on climate change."
—US Rep. Jim McGovern, member of Congress (D-MA)

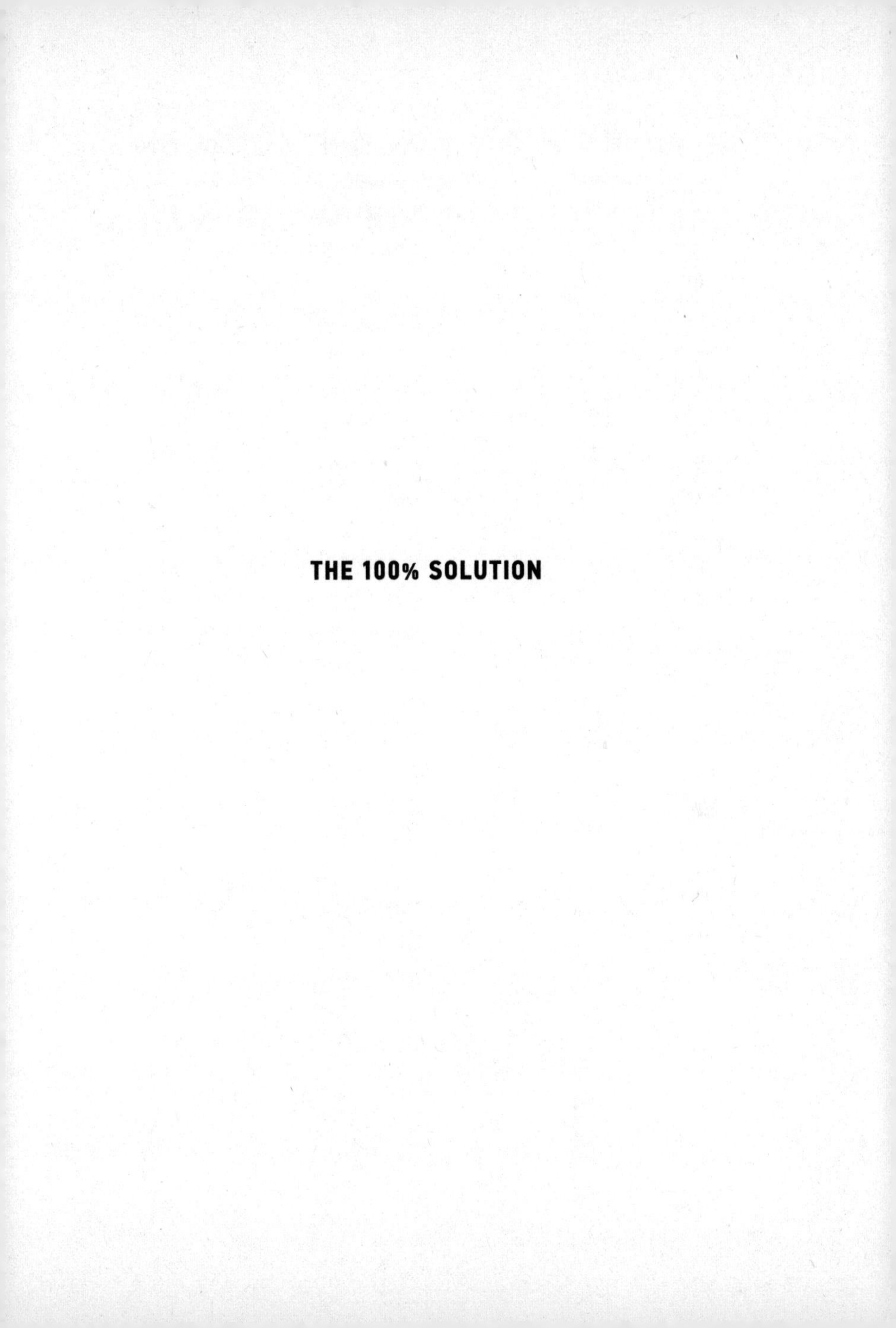

THE 100% SOLUTION

THE 100% SOLUTION

A PLAN FOR SOLVING CLIMATE CHANGE

SOLOMON GOLDSTEIN-ROSE

ILLUSTRATIONS BY VIOLET KITCHEN

MELVILLE HOUSE

BROOKLYN • LONDON

THE 100% SOLUTION

First Melville House Printing: March 2020

Melville House Publishing
46 John Street
Brooklyn, NY 11201

and

Melville House UK
Suite 2000
16/18 Woodford Road
London E7 0HA

mhpbooks.com
@melvillehouse

ISBN: 978-1-61219-838-5
ISBN: 978-1-61219-839-2 (eBook)

Library of Congress Control Number: 2019950168

Designed by Beste M. Doğan

Printed in the United States of America
1 3 5 7 9 10 8 6 4 2

A catalog record for this book is available from the Library of Congress

To my newlywed spouse Sophia Normark,
the only other absolute dream in my life aside from
solving climate change.

CONTENTS

INTRODUCTION

I'm twenty-six years old, and I can't remember a time before I understood and worried about climate change. Most Millennials take for granted that climate change is a serious problem that will cause devastating floods, droughts, storm damage, illnesses, and large-scale human and animal displacement in our lifetimes. And we know that if we don't reverse the current trend of greenhouse gas emissions, the effects will reach a new order of magnitude within our children's lifetimes.

Solving climate change is more important for our future than tackling many other worthwhile causes, because so many issues—poverty, disease, immigration politics—cannot improve if climate change worsens.

Like most people of my generation, I want to live in a safe world and pass on a better one to my children. Throughout my life, I've asked myself what I could do to help solve climate change. This drew me to study engineering, and later public policy; spurred me to run for Massachusetts state legislature right after college and get elected at age twenty-two on a climate-focused platform; drove me to spend thousands of hours reaching out to energy system experts, professors, startups, and fellow activists to understand the ins and outs of greenhouse gas emissions and climate change solutions; and prompted my recent shift to full-time climate change work.

This book is a product of that work. Along the way, I realized that most people, including me, tend to take on pieces of the problem that seem achievable, but don't understand whether or how those pieces might contribute to a solution that addresses the full picture. Perhaps most significantly, few people take into account the importance of developing countries, which emit two-thirds of global greenhouse gases each year.

Most of the specific ideas here have been written about before by other experts. The point of this book is to tie it all together into a framework that gives us a comprehensive perspective on what is needed so we can be more focused and effective in our advocacy.

Despite so much amazing work done over the years, the world is still nowhere near on track to solve climate change. The problem is not only worsening, it's worsening at an accelerating pace as we continue to add greenhouse gases to the atmosphere each year. The efforts of global agreements, national and state-level advocacy, and corporate promises combined have all failed to set us on a path that could add up to a complete reversal of greenhouse gas emissions in the timeframe laid out by the latest science—a 100% solution.

Part of our failure to act is because people haven't been totally sure what action is truly needed. For example, the school climate strike movement that youth leader Greta Thunberg is heading has repeatedly put out statements calling for policy-makers to "act" but not specifying what exactly they should do. Sometimes they've specifically said that it's up to the adults to figure that out. Meanwhile, legislators, who generally have equally incomplete understandings of energy and agricultural systems, push for incremental steps they see as politically viable and "in the right direction" without any sense of whether those steps could possibly add up to solving the problem globally. The US Millennial–led Sunrise Movement and related groups have called for a World War II–scale mobilization, which is an apt analogy for the general scale needed, but, within that broad vision, what *exactly* is needed? What constitutes "enough" to solve climate change? Without knowing the answer, we have no metric against which to judge political proposals or with which to guide how we direct our efforts.

This book has two goals. The first is to demonstrate how much bigger we need to be thinking if we're going to actually meet the consensus goal of climate scientists and the United Nations—reducing atmospheric greenhouse gas levels by the year 2050. The second goal is to lay out a specific framework, which is both technically and politically viable, that actually does add up to a 100% solution to climate change. My hope is that, with a specific blueprint to rally around, climate movements can focus their efforts and achieve more concrete results.

To be sure, small-scale actions will be necessary—someone has to carry out every little piece of the pillars I present here—but to be confident that those efforts will all add up in time, someone also has to be thinking about the comprehensive picture. That's the missing piece right now, and I hope this book will empower activists and the presidents and prime ministers they elect to be those comprehensive thinkers.

This book's framework centers on five pillars of action that add up to a 100% solution:

1. Deploy clean electricity generation.
2. Electrify equipment that can be electrified.
3. Create synthesized carbon-neutral fuels for equipment that can't be electrified or isn't electrified by 2050.
4. Implement various non-energy shifts, especially in agriculture.
5. Make up for the remaining emissions and get to negative emissions using sequestration.

This framework is compatible with various particular visions, proposals, and bills. In the United States for example, if Democrats who support a Green New Deal win enough Congressional seats and the presidency in the 2020 elections, that program would have a chance of achieving a 100% solution to climate change—so long as its specifics, when they are fleshed out, meet all the criteria of this framework. Or Republican Senator Lamar Alexander's suggestion for a New Manhattan Project and similar proposals that might come from members of Congress could provide a 100% solution if fleshed out to meet all the same criteria. Other proposals that presidential candidates might put forward could do the same in slightly different ways.

Or perhaps the United States will fail to take sufficient action and the president of China or a coalition of European and Asian countries will carry out the necessary projects instead. Or perhaps several political groups will propose different solutions that each add up to 100% but come with different benefits or costs. There are more efficient and less efficient ways to get to a full solution, but what matters most is that we achieve that full solution by 2050.

Despite a lot of implied and outright cynicism, it *is* possible to solve climate change fully by 2050. Yet, currently no proposal is on the table that specifically addresses the full scope of solutions needed, in part because there hasn't been a common frame of reference for what amounts to "enough."

Without key leaders taking a comprehensive view, we may not realize the need for certain tactics or the urgency of others. With a comprehensive view, the strategy becomes clearer, and we may find that the solutions to climate change that can actually add up are in fact more politically viable than the scattered actions we've attempted up to this point.

Indeed, the 100% solution laid out in this framework could be accomplished without personal or lifestyle sacrifices, without slow-moving international cooperation, and without compromising global economic development. Focused on innovation, and bolstered by policy wherever possible, this framework requires only the initiative of a small number of key entities to carry out the minimum required steps. Let's make it happen.

PART 1

SOLUTIONS MUST ADD UP

1

THE FULL SCALE OF THE PROBLEM

WE CAN'T HALF-SOLVE CLIMATE CHANGE

What does it mean to "solve" climate change? It is a problem that already impacts humans across the world, and it will continue to hurt today's young generations and the next few generations—as well as the planet's natural places and other species—even if we achieve the most ambitious targets ever set for "solutions." So, solving the problem can't mean avoiding any impacts; we've already failed at that.

But it also can't mean simply adapting to the new realities of a seriously warmed earth, because scientists tell us that we will reach a threshold of warming in the not-too-distant future beyond which the effects will be unacceptable or impossible to adapt to.[1] Whether disaster comes in the form of billions of people displaced by steadily rising sea levels or in the form of the possible sudden "tipping points" that could alter climate systems in some or all of the globe even more dramatically, we don't want to reach that point.

"Solving" therefore means limiting impacts to the moderately severe level that we can reasonably predict and adapt to, and which might eventually go away.

The interesting thing about climate change, which makes it different from most other global issues, is that we can't expect any lessening of impacts on the way to a full solution. Most social problems in the world, such as poverty or diseases, cause a certain amount of harm each year, and if we make a little progress on a given issue, it causes a little less harm the following year. Climate change is different. The impacts are caused by higher average global temperatures, which are driven by higher-than-average levels of greenhouse gases in the atmosphere. Reducing *emissions* is not enough to avoid, or even reduce, climate change effects—reducing the *amount* of CO_2 in the atmosphere is the only way to do that, which requires totally eliminating emissions and then removing some portion of the CO_2 that's already in the atmosphere. (We will discuss various methods for taking CO_2 out of the atmosphere later on.)

Eventually, we can return greenhouse gas levels, and therefore temperatures, to historic averages and be "back to normal." But the timeframe in which we do so is important, as "solving" climate change means ensuring that the *maximum* impacts—and any permanent changes they will cause—aren't impossible to adapt to. As temperature and impacts continue increasing until atmospheric CO_2 levels start decreasing, achieving the start of that decrease sooner rather than later is key in limiting the overall consequences of climate change.

For many years, political leaders and scientists worked under the rough estimate that limiting warming to at most two degrees Celsius would keep those overall consequences at manageable levels.[2] Even then, some cost-benefit models suggested 3.5 degrees or other targets as the standard, and various talking heads pretended the problem didn't exist at all.[3] More recently, as early climate change impacts have started to raise awareness of the issue, consensus about the importance of climate change has grown. In late 2018, the international scientific community—working through the Intergovernmental Panel on Climate Change (IPCC)—finally quantified what the world can expect with two degrees of warming compared to 1.5 degrees of warming, and concluded that 1.5 degrees should absolutely be the goal.[4] Most advocacy organizations and political leaders

Reducing emissions is easing up on the accelerator while still pressing it.
Eliminating emissions is taking our feet off the accelerator entirely.
Reducing atmospheric CO_2 levels is stepping on the brake.

have now adopted rhetoric consistent with the 1.5-degree target. This book will take that goal as a given and examine how to achieve it.

To do so, scientists tell us we need greenhouse gas levels in the atmosphere to start declining at a significant rate by around 2050, with as few emissions as possible between now and then. The moment we switch from emissions (adding greenhouse gases to the atmosphere) to "negative emissions" or "sequestration" (removing greenhouse gases from the atmosphere) is the moment at which impacts (perhaps delayed a few decades as temperature adjusts) start getting better rather than worse, with the promise of an eventual return to historic climate patterns. With the condition that any solutions that get us to that goal must be permanent, we'll call that turning point "solving" climate change.

There's a lot of rhetoric about "moving in the right direction" on climate change. But because of its difference from other issues (impacts being caused not by each year's emissions but by cumulative emissions until we start removing them from the atmosphere), there's not really such a thing as "moving in the right direction." *Climate change impacts get exponentially worse until we solve the problem 100%*. That's why it is so much scarier and more urgent than other problems, especially for today's young generations who, along with our children, would see the disastrous consequences of failing to achieve a 100% solution.

And that's why we must establish the timeframe: a 100% solution to climate change doesn't simply mean reducing emissions a bit each year until one hundred or two hundred years from now the whole world is carbon neutral. By then, catastrophic climate tipping points might have already disrupted the fabric of human civilization, and intense "normal" impacts would certainly have caused hundreds of millions or billions of deaths. "Solving" climate change—a 100% solution—*must* mean achieving negative emissions by 2050.

We can see how some strategies might be effective over the span of one hundred or two hundred years, but not within thirty. To achieve solutions, we'll need to think in this time-based mindset.

WHOA, THIRTY YEARS—CAN WE DO THAT?

Step back for a moment. We have just acknowledged that the only way to avert the most disastrous impacts of climate change is to totally eliminate emissions and start removing CO_2 from the atmosphere by 2050. That's a tall order.

It's such a tall order that many people think—consciously or subconsciously—that it can't be done. And historical experience weighs heavily on activists, policymakers, engineers, and others. Those who have lived to see climate change develop to this point may understandably think: "It's taken so much effort just to get pollution reduced a little, we can never make it go that much faster." Or perhaps: "We've been working on this issue for so many decades and yet it's still getting worse; we're probably doomed." Or a virtuous line of thinking: "We might be doomed, but I'll make sure I at least do my part and hope others do the same." Even the last of these does not commit to or contribute to actually solving the problem. Today's young people aren't looking for virtue; we're looking for results.

Many people, including the most passionate activists, seem to take for granted that we won't achieve a 100% climate change solution by the 2050 deadline that scientists have given us. So, naturally, they either disengage, assuming all is hopeless or that they personally can't affect the outcome, or they turn to "doing what they can."

For engineers, this means making modest progress on a certain kind of solar cell or battery, for instance. For activists and policymakers, this means pursuing state-level mandates for increased percentages of renewable electricity and public charging stations for electric cars. For corporations, it means well-publicized efforts to reduce water and energy use to "do our part" for the environment. And for many people, it means taking shorter showers and biking instead of driving.

The idea of "doing what we can" is dangerous when it comes to climate change because it implicitly accepts that the maximum viable action is less than the minimum needed action.

If advocates focus on "what we can" achieve right now, even if all of their efforts were successful, the results will fall short—as they have for many decades.

Instead, we need to approach climate change from the perspective of "what needs to be done," and come hell or high water—the latter may well come literally—find a way to make it happen. This is what younger people understand, and why the movements getting new levels of attention around the world are youth movements demanding action to truly solve climate change.

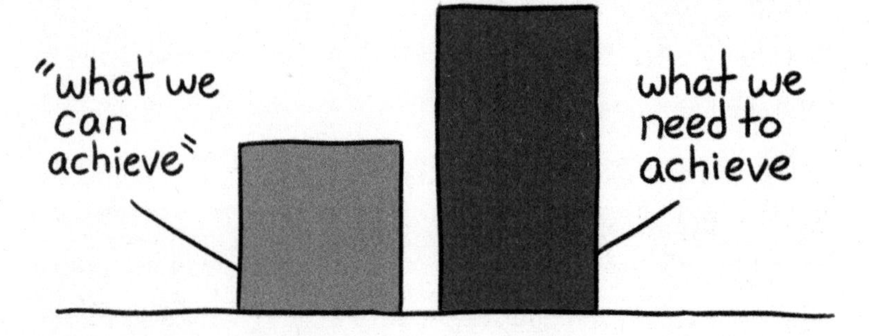

MILLENNIAL THINKING

The daunting need for a 100% solution may be where young people's perspective is most needed. When examining a problem, older people often consider the issues in the context of their life experience: how an issue came to be, how similar issues have been addressed in the past, and so on. Young people have yet to acquire most of that perspective. Sure, we can learn from historical examples, but we can also take issues solely in the context of where they are right now and where they might go from here.

Looking backward in time is often an essential contribution to solving a problem. But in the case of technology, for instance, young people's tendency to only look forward becomes particularly useful. Political rhetoric may return repeatedly to the same ideas, but technology keeps moving in one direction.

No one has tried to return society to horse-drawn carriages. And legislatures around the world are learning that they'd better license and regulate new ride-sharing apps because no city is going back to traditional taxi dominance.

The only way to avert the most disastrous impacts of climate change is to totally eliminate emissions and start removing CO_2 from the atmosphere by 2050. That's a tall order.

A forward-looking perspective is all the more important when dealing with a challenge unprecedented in human history, because there *is no* comparable situation on which to base our thinking.

In the 1970s, advocacy efforts passed policies to clean up smog in US cities. The Clean Air Act took a lot of political will and regulatory enforcement, and the environmental activism and policy leadership involved was hard and slow. It involved top-down mandates and slightly burdened some companies with extra costs, and therefore it was a hard-won victory. People who lived through the decades in which smog was dramatically cleaned up might compare that effort to today's need to clean up greenhouse gas pollution. It might be easy to assume that addressing climate change will also involve a slow, mandate-based transformation over many decades, and that incremental actions are all we can hope for today—even if we know such actions won't add up in time to solve climate change.

Young people look at the numbers scientists have given us our whole lives and say, "Whoa, we need to find a plan that can make this transition happen in thirty years!" There's no question of whether this is likely or not given historical examples. We know that our generation's future and our children's future depend on finding a way to achieve a full solution to climate change in thirty years. We acknowledge that as the baseline and think forward: How do we make that happen?

THIRTY-YEAR SOLUTIONS

For emissions to be eliminated and sequestration to begin by 2050, various tactics will have to be employed that add up to a 100% solution. Most crucially, all tactics that we're going to *rely* on must be able to roll out at full scale between now and then. (Tactics that would be useful but that we're not relying on in the minimum calculation of adding up to a 100% solution do not necessarily need to achieve full scale by 2050.)

"Full scale" means a tactic has been adopted everywhere we're relying on it being adopted. For example, if a mandate for capturing CO_2 from the exhaust of steel factories is a tactic our strategy relies on, then that mandate must be passed in almost every country that manufactures steel—close enough to all of them that the tiny remaining steel emissions could be made up for by sequestration (which is almost always more expensive than eliminating emissions in the first place, and is therefore the "last mile," not "first mile," of emissions reductions). To include such a mandate in a strategy that adds up, we have to be confident that it could in fact be politically viable in enough countries.

Likewise, relying on a particular technology—such as electric heat pumps displacing methane ("natural gas") and oil for home heating—means that we must believe it is or can become cheap enough to be adopted pretty much everywhere in thirty years. Obviously, if it is twice as expensive as the fossil fuel alternative, not many people will choose to make the switch—and not many countries will be able to garner the political will in such a short timeframe to mandate it.

There are several key questions for evaluating tactics and strategies that could add up to a 100% solution to climate change:

- Is it or can it be cheaper than, or about the same cost as, the current system? If a clean option is significantly more expensive, it is too unlikely for policy incentives or mandates to be viable in enough countries to rely on that option as part of a 100% strategy.
- Can it physically scale up in time? Are there enough raw

materials? Can the supply chain expand rapidly enough to manufacture/distribute/etc.? Will there be a workforce to build or run or teach about it? How much land area does it require?

- How many people have to make a proactive choice to adopt it? A solution implemented at the governmental or wholesale level is much more likely to reach full scale in thirty years than a solution that requires every family in the world to make an individual decision.
- Does it require significant changes to anyone's lifestyle? While solutions that focus on behavior change may have a chance of significantly shifting consumption trends over many decades, at the global scale, these solutions won't add up in a major way in thirty years (especially when so many people have serious financial limits). Personal lifestyle changes fall in the "cherry on top" category of tactics to try for, but not rely on.

In constructing a framework for a 100% solution, all the tactics we plan to rely on must have satisfactory answers to these questions. Most solutions must be cheaper than the current system to scale quickly. If they are significantly cheaper, they will roll out faster—a good example of this is how methane power plants rapidly replaced coal power plants in huge swaths of the United States as fracking made methane a much cheaper fuel in the beginning of this century.[5] Coal workers didn't like being outcompeted, environmentalists hated fracking, and politicians promised to reverse the decline of coal, but it kept happening because methane was so cheap due to the new drilling technology.

Most solutions must require no or minimal changes to people's lifestyles. A good example of how lifestyle-changing solutions can fail to be adopted even if they have other benefits is the case of clean cookstove deployment. A massive effort was made in the last two decades to give families in developing countries electric or propane cookstoves that don't pollute their homes or cause burns. Follow-up studies found that many families in India continued using their old, more dangerous wood-based cooking equipment because of its cultural and traditional value, ignoring the healthier and cleaner stoves they were given.[6]

CURRENT RHETORIC DOESN'T ADD UP

We need a framework to guide activism and policy efforts—one that actually adds up to a 100% solution: negative emissions by 2050. So far, political discourse has been filled with incrementalist steps (for example, improving car engine efficiency) or with rhetoric that focuses too narrowly on one tactic or set of tactics without concern for the comprehensive picture (for example, "leave fossil fuels in the ground" or "mandate 100% renewable electricity").

Youth movements such as the school climate strike have recently raised climate change much higher on the political agenda, and they have highlighted the urgency and scale in a way no one else has been able to. But these movements still don't have a specific plan or proposal that would achieve a 100% solution. The strikers demand "action," and, indeed, our policy leaders should heed their call and get to work. When they do, we will need to have a rubric by which to judge whether policy actions are significant enough.

In the United States, the Green New Deal was popularized by the youth-led Sunrise Movement and its transformational aspirations—"WWII-scale" mobilization—are around the right scale, but again it doesn't include specifics on how we're going to achieve all the aspects of this transformation. Similarly, some 2020 presidential candidates have proposed excellent ideas, but always with at least some piece missing or without a clear commitment to the full scale of action needed. The lack of full solution–oriented proposals may represent a deeper sentiment—that we don't *need* to analyze how things add up, but simply need to start taking actions that reduce emissions. This might come from a line that gets repeated often by climate thought leaders and has pervaded much of the activist community: "We have all the technology we need; we just need political will."

If that were true, then a demand to simply "act" would make a lot of sense, as would a policy mandate to simply stop using fossil fuels.

The problem is that we *don't* have a lot of the technology we need—for more than half of global emissions, the theoretical technologies that may exist in a lab or demonstration project somewhere are far too expensive to roll out economically by 2050 and usually too expensive even to make policy mandates a viable option. Those technologies will all have to improve rapidly to enable a transition by 2050, and they will have to improve *dramatically* if the majority of the world that is poor, developing, and reliant on cheap energy is to adopt them. Indeed, the idea that political will alone can solve the problem is incredibly industrialized-country-centric.

2

GLOBAL EMISSIONS, GLOBAL SOLUTIONS

10% VS 100% SCENARIOS

The dialogue about having all the technology we need and only lacking political will arose in the 1990s and early 2000s. It seems to have been Al Gore who popularized the line to emphasize that solar and other such options were viable right now and that we could choose to adopt them if we were willing to pay a little extra now to avoid massive costs from climate change impacts later.[7]

It's a useful way to emphasize our ability to *start* reducing emissions. The problem is that it tends to imply we currently have the ability to simply "choose" to eliminate *all* emissions. Even in the realm most activism focuses on—electricity generation—we can see this is not true. Yes, we can choose to start replacing coal and methane ("natural gas") power plants with solar and wind electricity generation. An electric utility company can build a bunch of solar farms and they'll run their methane power plants a little less. Since the early 2000s, solar installation costs have dropped enough that they might even save money, at least with the income from state and federal subsidies. Indeed, some states have driven this process along to the extent that solar now comprises about 10% of their electricity generation—so, political will can cause some percent of methane to be replaced with solar. But if we pushed that to 100% and shut the methane plants down entirely, we wouldn't have electricity at night. The 100% scenario isn't the same, engineering-wise, as the 10% scenario.

A longer-sighted political argument would acknowledge that the anti–climate action politicians are *partly* right: some technologies—such as batteries to store solar energy for a full nighttime—are too expensive to adopt right now. Rather than pretending that everything is cheap enough already, or succumbing to the defeatist (or denialist) argument that we should wait for technology improvements to come along before taking any meaningful climate change action, pro-action politicians should emphasize the need to *create* those improvements in technology so we can mandate solutions where necessary, and see other technologies get adopted naturally as they become the cheapest option in the market.

Sadly, this argument was largely ignored for many years because of the increase in partisan polarization—at least in the United States—that made each side feel like it had to totally deny all arguments the other side made. This polarized atmosphere has made it seem that if climate deniers were right in one tiny piece of their rhetoric, the whole pro-action argument would fall apart, which of course is not the case. Recently, many pro-action politicians have gotten closer to the mark, acknowledging the need for technology improvement alongside short-term acceleration of deployment. In general,

> **Pro-action politicians should emphasize the need to *create* those improvements in technology so we can mandate solutions where necessary, and see other technologies get adopted naturally as they become the cheapest option in the market.**

heightened attention to climate change has started to reduce polarization around the issue and pushed a larger portion of politicians toward practical strategies (see Chapter 11).

To achieve a 100% solution, it is not enough to simply start taking various actions that reduce emissions. We need to analyze how things add up in the comprehensive picture and strategize accordingly. And technology improvement is essential to any strategy.

INDUSTRIALIZED COUNTRIES

Technology improvements that lower the cost of non-polluting options would make it much more politically viable to mandate their adoption. It is theoretically possible that once clean technologies are only slightly more expensive than their polluting counterparts, industrialized countries could mandate solutions that cost a little extra and eliminate nearly all their emissions. One study that modeled strategies for the United States to achieve 80% emissions reductions by 2050 showed the cost would only be around 1% of GDP.[8] That's politically hard, but not actually unreasonable. Perhaps the remaining emissions would be minor enough that they could be sequestered, and industrialized countries would be carbon-neutral.

This isn't a likely pathway—it relies on political will being strong enough in the next few years to sell cost increases to an agitated public. And it might turn out that some specific technologies are simply too expensive to pay for, even if the political will exists. However, we shouldn't rule it out. For industrialized countries that have the money in individuals' pockets, companies' investments, and governments' budgets, at least a large share of emissions *could* be addressed simply "by having the political will."

DON'T FORGET INDIA

But rhetoric that implies the whole solution to climate change could come from such incentives or mandates ignores the key principle that any solution must be viable in developing countries as well. Although most cumulative emissions since the Industrial Revolution have come from those countries that are now the wealthiest industrialized nations, two-thirds of *current* emissions—the emissions that will make the difference between achieving and failing to achieve a 100% solution — come from developing countries, which are now undergoing that same process of industrialization.[9]

Unlike industrialized countries, where mandates and subsidies run into mainly political problems (at least once clean technology becomes only slightly more expensive than current options), developing countries often lack the physical ability to deploy more expensive solutions, whether they would be paid for by individuals, companies, or government bodies. Indian political leaders would love to clean up the air pollution that plagues their cities and reduce the threat of climate impacts that menace their coasts, but their country doesn't have the extra money to simply make that choice. And India is representative of many developing countries around the world.

Huge swaths of Asia and Africa and parts of Latin America—together containing more than half the world's people—are climbing rapidly out of poverty. For this more-than-half of humanity, industry is growing, medicine is becoming more accessible, education is reaching most of the population, and birth rates are falling.[10] Other countries have gone through similar development and seen these dramatic improvements in quality of life, but remain much less wealthy than the richest industrialized countries, falling in the same "middle income" category as India.

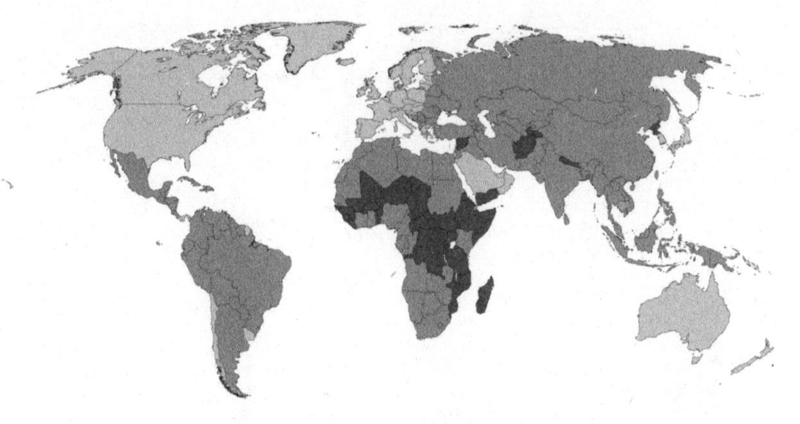

Countries according to the World Bank's categories of high income, middle income, and low income.[11]

Many other countries are still in the earliest stages of poverty reduction, and few people have significant access to energy. Those "lowest income" countries are the places where population is growing fastest, but energy use (and food consumption) per person is still so low that they aren't contributing significantly to climate change. The larger contributor is the set of "middle income" developing countries epitomized by rural India, which is in the second phase of development, where population growth has started to slow but where *access* to energy has started to grow rapidly, causing major increases in greenhouse gas emissions.

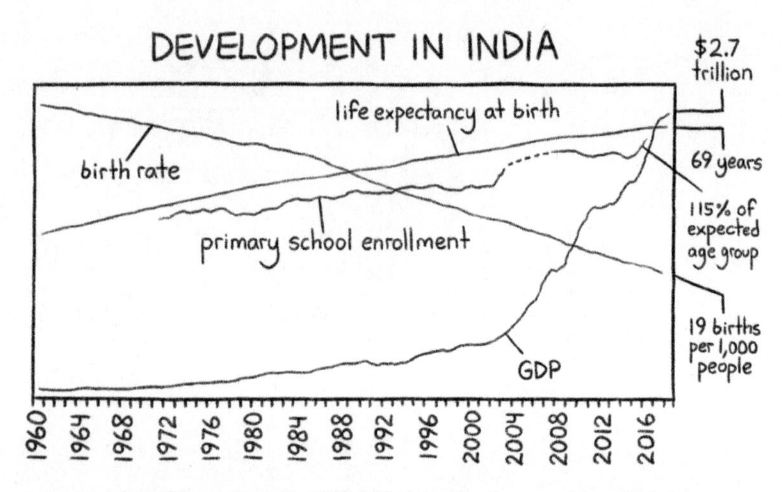

Indicators of development in India, 1960–2016 (dotted line for years from which data are not available); all four lines are on y axes starting at zero. Primary school enrollment exceeds 100% because students older than the official primary school age range are also enrolled. [12]

In these countries, access to electricity for safe lighting, access to home heating and air conditioning, the ability to get around quickly, and industry that provides jobs and raises incomes are all being powered by fossil fuels. This is because fossil fuels are cheap. In poor parts of the world, people can only access energy when that energy is cheap. Paying more for clean options is impossible. For the same reason, policy mandates that achieve cleaner systems at extra costs are impossible. This is why policies that simply mandate adoption of current, more expensive clean options are unlikely to happen in most of the developing world. Fossil fuels are lifting people out of poverty—as they once did in now-industrialized nations.

You can't ask a poor farming community in India to forgo a coal power plant and remain in poverty; you have to have something cheaper. When political discourse in industrialized countries focuses on policy mandates that would increase costs, even slightly, policymakers and activists are forgetting India.

Given the huge share of global emissions that come from this middle category of countries, that's a serious problem. Currently, developing countries emit about two-thirds of the world's greenhouse gases each year.[13] Despite having significantly lower emissions per person, a large majority of the people in the world live in developing countries, not in

Developing countries emit about two-thirds of the world's greenhouse gases each year . . . You can't ask a poor farming community in India to forgo a coal power plant and remain in poverty; you have to have something cheaper.

countries with extensive already-established energy infrastructure. In developing countries, such infrastructure (currently almost all powered by fossil fuels) has expanded rapidly, first in China in the last few decades and now in India and elsewhere. Because they are the countries where energy access is growing, developing countries account for not only two-thirds of current emissions but also almost the entire *increase* in emissions under the business-as-usual models between now and 2050.[14] For example, a recent International Energy Agency report noted that newly introduced air conditioning alone is projected to increase electricity demand massively in developing countries by 2050—at a deployment rate of roughly ten AC units every second.[15] In business-as-usual models, 70% of total cumulative emissions from 2020–2050 would come from developing countries. Therefore, solutions to climate change must first and foremost be practical in these countries.

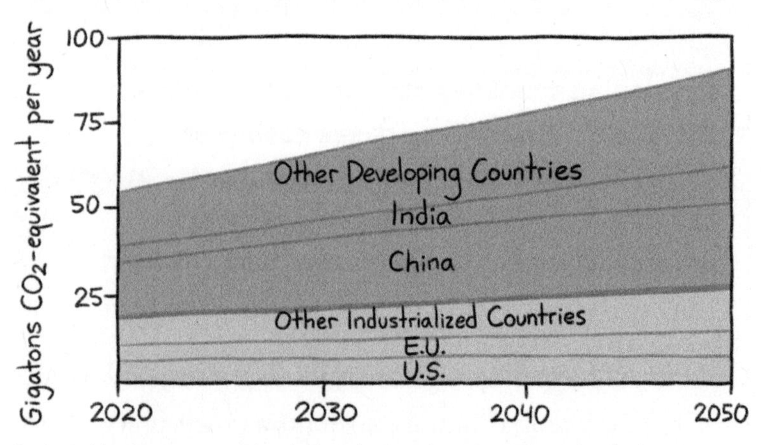

Projected business-as-usual emissions from various countries—developing countries contribute 70% of the cumulative total from 2020–2050.[16]

Because developing countries can generally make the necessary transitions only if clean options are cheaper, the mandate-based policy discourse of industrialized country activists and policy leaders applies to only a third of the world's emissions problem. (Of course, some parts of "developing" countries are industrialized, and some parts of "industrialized" countries are still developing, but the rough proportions of two-thirds/one-third serve as an average across the world.)

> **The United States will still be hit with the full impacts of climate change—even if industrialized countries successfully decarbonize—unless the entire world eliminates emissions in time.**

Maybe people assume that developing countries' emissions aren't so significant. True, developing countries currently use less energy per person, but a majority of the world's existing population lives in these countries, and their usage per person is increasing massively every year.

Or maybe people are stuck in that mentality of "doing our part" without considering that some ways that "we" might be able to do our part would not work for most of the world. However and whenever the United States and other industrialized countries decarbonize, it will not matter much unless the rest of the world does so as well. US activists and political leaders need to remember that the United States will still be hit with the full impacts of climate change—even if industrialized countries successfully decarbonize—unless the entire world eliminates emissions in time.

ARE CLEAN OPTIONS REALLY MORE EXPENSIVE RIGHT NOW?

Some climate activist and renewable energy company rhetoric implies that clean options are already cheaper than fossil fuels. This is true in some cases—for example, in a region with abundant potential for hydropower, that is usually the cheapest option for electricity generation.[17] Countries with huge hydropower potential have indeed exploited it to a large degree, sometimes powering almost 100% of the nation's electricity. Brazil, for example, is a developing country that powers about three-quarters of its electricity using hydro.[18] That's because naturally, the cheapest electricity generation option is the one that is deployed most. By the same logic, we can see that in most parts of the world without that kind of hydro potential, fossil fuels—not wind or solar—are the cheapest options.

In the United States, as methane became very cheap in the last two decades, methane power plants rapidly replaced coal power plants. The fact that India and China are still building large amounts of coal electricity generation demonstrates that in those regions, coal is indeed the cheapest energy source.[19] There are some cases where this is starting to change (such as with electric cars, which are cheaper to operate than gas cars but not yet cheaper to purchase), but by and large, the continued massive expansion of fossil fuel energy around the world shows that clean options are still more expensive.

A confusing factor is that *when* they are producing electricity, solar and wind can be cheaper than coal and methane, which has led to many misleading reports that renewables are "already cheaper" than fossil electricity. But because they don't produce electricity most of the time (in fact sometimes output drops for days at a time, a duration no battery system can yet come near to sustaining economically), the total costs of electric grids heavy in solar and wind are higher than those of grids heavy in fossil fuels. That's why developing countries are not deploying renewables more rapidly.

INNOVATION IS NEEDED

The solution, of course, is that industrialized countries must use their greater financial flexibility and technological expertise to bring the cost of most clean technologies down to levels roughly on par with, or lower than, current emitting systems. This is the only way clean options can scale up in the developing world fast enough to eliminate nearly 100% of emissions by 2050.

There are several ways to make clean options cheaper: simply scaling up the manufacturing of clean options can bring down costs through more efficient business practices and supply chains. In some cases, new basic technology can improve the cost of an existing product—for example, a new electrode material for a lithium-ion battery. In other cases, brand-new designs for the whole product can become the key innovation—for example, a liquid flow battery that has a different structure from a lithium-ion battery entirely. In many cases, technologies (both for entire new products and for components or new manufacturing systems) are floundering at one of these early stages. "Innovation" is used throughout this book to refer to any and all direct efforts to move them along toward widespread commercial deployment.

The interesting part of innovation-based solutions is what they mean for industrialized countries' own transitions: If clean options can become cheaper than dirty options virtually everywhere in the developing world, they will be cheaper in industrialized countries as well. Once that happens, it will eliminate most of the need for policy mandates (at least for energy emissions) both in developing and in industrialized countries: the cheaper clean technologies will outcompete polluting technologies naturally. The cheaper the clean tech can become, the faster the transition will happen.

Some mandates or incentives in industrialized countries may well be required to bring those technology costs down in the first place. For example, California's Low Carbon Fuel

Standard is already helping to enable commercial development of carbon-neutral fuels that otherwise would be slower to get to market. But by the end of the transition, if it is to go fast enough, policy approaches won't be the main direct driver of emissions reductions in either developing or industrialized regions.

When current rhetoric says "we have all the technology we need, we just need political will," it overlooks the fact that most clean options are still far too expensive to mandate, let alone scale on their own, within thirty years, and it focuses too narrowly on industrialized countries "doing their part," ignoring the fact that political will can never overcome basic economic realities in the developing countries that account for two-thirds of the world's emissions.

It is time to leave that thinking behind and get into the mindset of a 100% solution. We need tech improvement to make clean options cheaper, to make *all* emissions—including those from the developing world—reach near-zero before 2050. Then we will need affordable sequestration methods to reach negative emissions by 2050 to save today's young generations and our children and grandchildren from the worst impacts of climate change.

3

ELIMINATING *ALL* EMISSIONS

EFFICIENCY TACTICS ARE NOT SOLUTIONS

Think of the example given in the previous chapter of building solar panels and running methane power plants a bit less. It worked for the 10% scenario, but it didn't include a plan for the 100% scenario accounting for when electricity wouldn't be generated at night. Different measures, such as mixing in other electricity generation technologies or adding long-term electricity storage systems, would be needed to preserve a steady flow of electricity in the 100% scenario. But those measures won't be pursued if our efforts remain fixated on the 10% solution.

The mindset that all 10% solutions could be pushed to become 100% solutions makes a lot of advocacy rhetoric unfocused right now. It implies that tactics that can bring immediate emissions reductions are the same tactics that need to be pushed further to achieve all emissions reductions, which is often not true. This conflation of 10% and 100% solutions might come partly from a combination of normal-issue thinking (reduce the cause of a problem a bit and that will reduce the impacts a bit) and subconscious cynicism (we'll probably fail at the 100% solution so I won't even try to think about the full picture, but will focus on a piece that I can understand or affect). A lot of efforts and proposals up to this point have focused mainly or entirely on tactics that can reduce emissions by some percentage economy-wide or in a given sector, but cannot, by definition, ever reduce them to zero.

We'll call these efforts "efficiency tactics." Some include efforts for literal efficiency: requiring cars to have better miles per gallon, or subsidizing consumers to buy more efficient oil boilers for heating homes and businesses. Some are indirectly in this "efficiency" category: some climate change proposals have called for efforts to expand education of women and girls and access to contraception in developing countries. These are worthy causes, but their climate change impact is limited to reducing the growth in population, a minor factor contributing to our current emissions trajectory (as noted before, the countries with the fastest population growth are those in the first stage of development, which will not see massive energy access expansion in the next thirty years).

Reducing population growth equates to reducing the 2050 demand for energy and food, but can never get near eliminating that demand. And incentivizing more efficient fossil fuel equipment, or replacing the first 10% of methane electricity generation with solar, means reducing emissions within a system that still causes emissions. What we need instead is to shift to systems that cause no net emissions.

Many efficiency tactics can be helpful in driving system shifts. For example, carbon pricing (charging a fee on fossil fuels and emissions-intensive goods) would itself reduce but not eliminate emissions, but it would also create markets for new clean energy technologies and accelerate the adoption of clean options (in fact, carbon pricing is probably the single best incentive policy to support and accelerate the implementation of this framework). Efficiency tactics can reduce the total cost and make the transition to non-emitting systems easier. It's even possible that in some countries, the only way to implement the needed physical changes is through a combination

CURRENT SYSTEM
(Fossil fuels)
CO_2

EFFICIENCY (Reduced demand)
CO_2

NEW SYSTEM (Electricity)

of many such efficiency measures. But they cannot themselves add up to 100% globally. Policy leadership and advocacy need to focus mostly on the minimum framework that constitutes "enough" to actually reach a 100% solution.

Here's another related example. Many activists emphasize the need for individuals and institutions to "do their part."

Turn off your lights when you're not in the room...

Insulate your house better...

Recycle more...

When everyone who is willing pools their individual behavior changes, these steps might all add up to a 10% solution. But these are only the most tangible sources of emissions from individual lifestyles. Aside from the fact that many, if not most, people will not be convinced to "sacrifice" and change their behavior to reduce emissions, even if they did we would be doing nothing about the many emissions that are inherently tied to larger systems. We must not confuse steps that put small dents in emissions—the 10% solutions—with system-changing steps that could add up to eliminating all emissions globally—the 100% solutions.

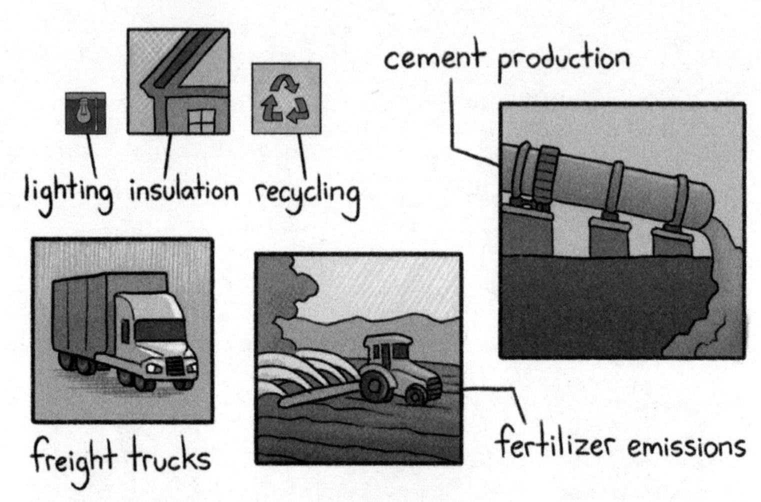

Size proportional to potential global emissions reductions.

Even the best advocacy efforts focused on 10% solutions may still take away effort or momentum from the 100% solutions that are really needed. And worse, some movements toward 10% solutions actually make it harder to later implement 100% solutions, by locking in infrastructure that can't accommodate key 100%–solution technology, or by creating a mindset that ignores the need for other tactics to reach 100%—a mindset that can be difficult to reverse in time.

> **We must not confuse steps that put small dents in emissions—the 10% solutions—with system-changing steps that could add up to eliminating all emissions globally—the 100% solutions.**

THE FULL PIE

A 100% solution has to deal with all the greenhouse gas emissions in the world.

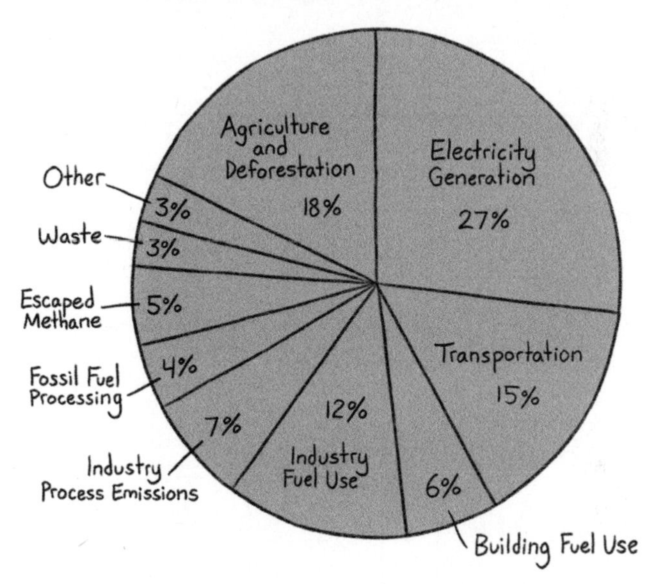

A 100% solution means eliminating all these emissions, and then starting to remove CO_2 from the atmosphere to get negative emissions. Current political discourse in industrialized countries focuses mostly on electricity generation, which is responsible for only about 25% of global emissions. Some efforts target transportation and building heating, which together add only about 20% more. In total, all the usual rhetoric addresses less than half of emissions.

GLOBAL SOURCES OF GREENHOUSE GAS EMISSIONS[20]

ELECTRICITY GENERATION
(INCLUDES DISTRICT HEAT PLANTS AS WELL)

Mostly coal-powered, lots of methane and some oil; centralized infrastructure, owned by public or private utilities and energy companies; lasts for many decades.

TRANSPORTATION

Virtually all oil-powered; mostly road travel (personal vehicles and freight trucks); decentralized infrastructure, owned by billions of people and companies; lasts ten to twenty years; ships ~2%, airplanes ~1.5% of total pie, owned by smaller number of entities.

BUILDING FUEL USE

Mostly methane and oil (furnaces) with some coal and wood (heat stoves); mostly for space heating, some water heating and cooking; decentralized infrastructure, owned by billions of people and companies; lasts fifteen to thirty-plus years.

INDUSTRY FUEL USE

Mostly coal, mostly for heat to drive processes; many subcategories, most significant are steel furnaces and cement kilns; semi-centralized infrastructure, owned by many companies, large portion in China.

INDUSTRY PROCESS EMISSIONS

Direct emissions from chemical reaction byproducts other than burning fuels (e.g., CO_2 off-gassed when limestone heated to make cement; CO_2 released when methane split to get hydrogen for chemical feedstock); centralized infrastructure, subset of industry fuel use ownership.

FOSSIL FUEL PROCESSING

Burning fuels to extract and move coal/oil/methane.

ESCAPED METHANE

From extraction and distribution leaks, incomplete combustion in compressor engines along methane pipelines, and extraction equipment; centrally-owned but physically spread out infrastructure.

WASTE

Organic waste in landfills produces methane and wastewater treatment plants produce several greenhouse gases; mostly-decentralized infrastructure.

OTHER

Processes that don't fit into other slices.

AGRICULTURE AND DEFORESTATION

Cutting and burning forests (and other ecosystems) releases CO_2 (7–10% of pie); livestock belch methane (~5%); fertilizer generates nitrous oxide emissions (4–5%); soil, residues, and equipment release other emissions.

If we want a 100% solution, we need proposals to address 100% of the pie. The less-discussed half of the pie includes many smaller slices that aren't even related to burning fossil fuels—non-energy direct emissions from certain industrial processes, waste, and more. The largest of these slices is agriculture, which includes emissions from deforestation and other changes to land.

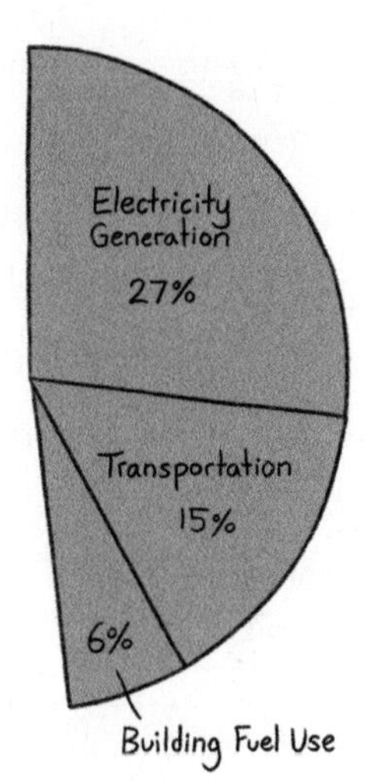

All the usual rhetoric addresses less than half of emissions.

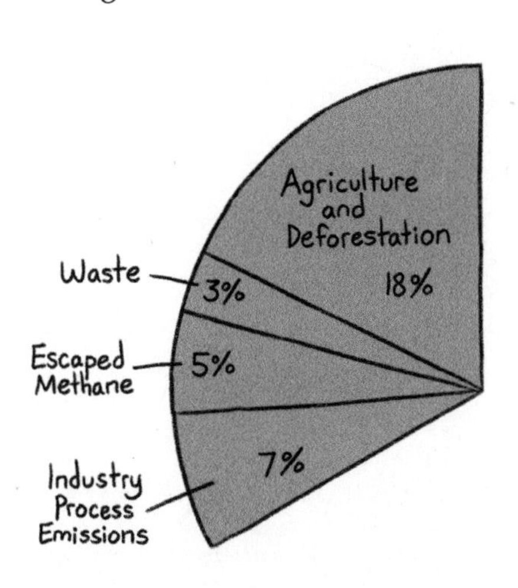

Non-energy emissions are about a third of the global total.

Some of the larger fossil fuel slices in the chart are composed of many different processes—most notably industry fuel use, which encompasses various activities with totally different kinds of equipment used to produce everything from steel and cement to paper and ammonia. None of these slices tend to be included in mainstream rhetoric, yet they need to be eliminated as well. Many of these specific sources of emissions will be discussed in more detail when we look at solutions to them in Chapters 6 and 7 (for industry energy, as well as transportation and heating) and Chapter 8 (for non-energy emissions from industry, agriculture, and wastes).

TECHNOLOGY DEVELOPMENT IS NEEDED

We can simply look around and see that there aren't widely deployed non-emitting technologies for many of these slices. Researchers and startups are working on most innovations needed, but a large portion are still in lab stages or are still far too expensive to implement, even with serious subsidies. For example, there isn't a technology readily available to power airplanes in a fully carbon-neutral way, even at a modestly higher cost than that of jet fuel.[21] In fact, the majority of the pie does not have currently affordable technology that could be mandated immediately to eliminate the slice in question. A 100% solution requires lowering the cost of technologies to address the remaining emissions.

Some of the technology required is a piece of equipment—for example, someone has to design and commercialize an electric motor to power long-distance ships. Some of the technology required is a process—for example, cement production currently off-gasses CO_2 when limestone is heated in a cement kiln, and someone has to design and implement a new process that reduces or eliminates those direct emissions. Some of the "technology" required is in the form of new practices—for

> **The majority of the pie does not have currently affordable technology that could be mandated immediately to eliminate the slice in question. A 100% solution requires lowering the cost of technologies to address the remaining emissions.**

example, someone will have to work with farmers to get most farms in the world to adopt better soil management systems (which relate to crop rotation, interspersing of crops and livestock, organic practices, and more) to reduce emissions from farmland soil, and to start sequestering CO_2 in soil, which there is significant potential for. And finally, some emissions—for example, direct emissions which are caused by fertilizer spread on fields (see Chapter 8)—probably can't be fully eliminated by 2050.

By 2050, we will need sequestration to compensate for those emissions that can't be fully eliminated. Sequestration, of course, is needed by definition to go from zero emissions to negative emissions.

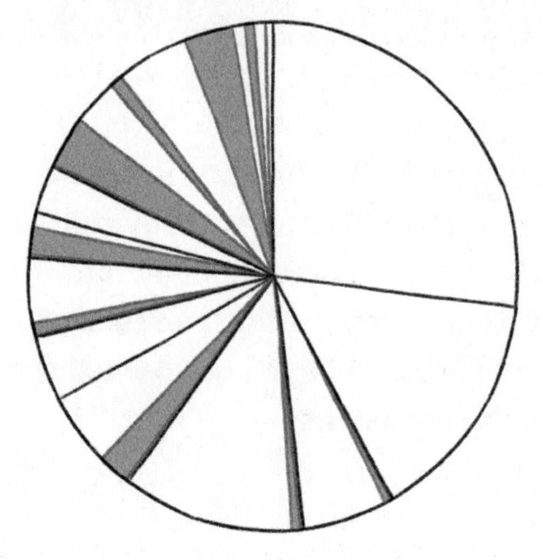

These slivers of emissions probably can't be eliminated by 2050.[22]

"Sequestration" means removing CO_2 that is already in the atmosphere and storing it (as pure CO_2 gas or as another chemical that it gets turned into) in some way that it will not be released again for a long time, if ever. CO_2 can be locked into the wood, soil, and other plant matter in forests; mixed into farm soils by microorganisms and crops; pumped as a gas into underground caverns; or chemically converted into plastics or other goods that lock the carbon inside their materials. These and other methods will be explained in Chapter 9.

Some of the ways to accomplish sequestration—especially soil management and replanting forests—are cheap or might even save farmers money, and they come with other benefits to ecosystems, outdoor recreation areas, and agricultural yields. But they can only sequester a certain amount of CO_2 each year—there's only so much land area where we can plant forests.

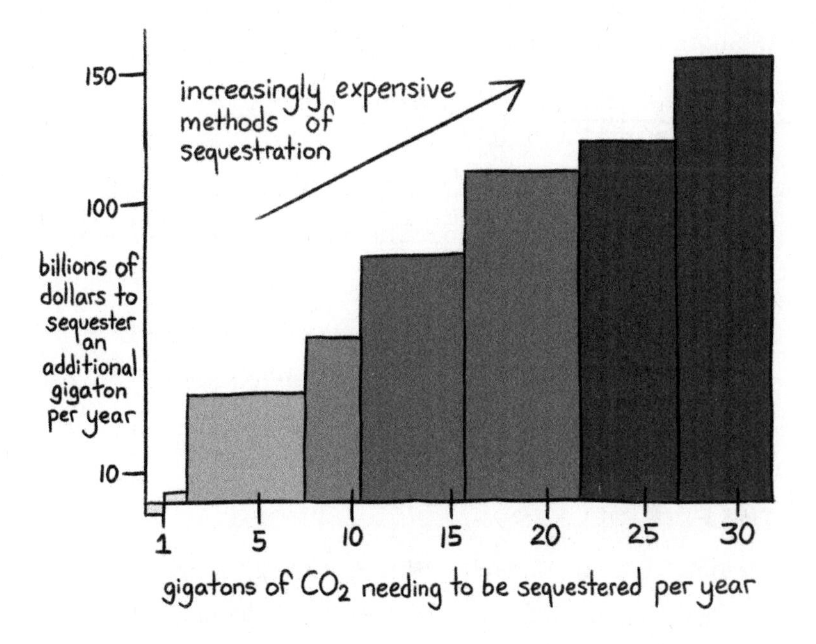

Cost to sequester each next ton of CO_2 increases significantly the more tons that are required because we will have to use more expensive methods—see Chapter 9.[23]

The more emissions remain in 2050, the more we will have to rely on higher-cost sequestration methods, such as capturing CO_2 directly from the air and pumping it into deep underground caverns. Doing so will require serious technology development and improvement to bring down costs to a reasonable level. Even then, governments will have to pay for this portion of sequestration outright, with no co-benefits. Hopefully that will be politically viable by the late 2040s. In the best-case scenario, the total cost worldwide might be in the hundreds of billions of dollars per year. If sequestration technology isn't improved—or more significantly, if other emissions aren't reduced to near zero—and the cost is in the trillions of dollars per year, it may not be politically feasible to pay for enough sequestration. Therefore, the work of not only reducing, but virtually eliminating, emissions is essential to bring the world to the point that negative emissions can ever be achieved. This will depend on innovation to improve the cost of various technologies.

THE FIVE PILLARS

We now have an understanding of the full scope of where emissions come from and the full, global scale of transformations needed. Efficiency tactics will not get us to negative emissions by 2050, nor will incrementalist policies that drive small- or moderate-scale adoption of clean technologies but have to be enacted one by one in every single country. The viability of a thirty-year transition to negative emissions depends on dramatic amounts of innovation to lower the cost of clean technologies and practices.

Putting all the pieces together, we can now see exactly what that innovation has to consist of.

The obvious culprit for climate change in most people's minds is electricity generation. This is indeed the area that offers the most immediate opportunities for switching to non-emitting systems. We'll call decarbonizing electricity generation Pillar 1, the first area of work that is essential to a 100% solution.

With lots of clean electricity available, we can then transition the rest of the energy system to clean options—many processes can be electrified, as is starting to happen with cars and home heating. As noted, electric ship engines could be designed and commercialized. Such efforts are the work of Pillar 2.

However, not everything in the energy system can be electrified—long-distance plane trips, for example, rely on such concentrated fuels that it is highly unlikely there will be affordable electric options by 2050. The same goes for various industrial processes. Furthermore, not every building will be renovated between now and 2050 to replace its existing furnace with electric heating. Therefore, this book's framework adds a piece often left out of popular rhetoric: synthesis of carbon-neutral fuels to substitute for fossil fuels in energy applications that can't be electrified by 2050, or aren't converted in time. This may include some amount of biomass-derived fuel, but mostly, assuming imperfect policy around the world, much of it will have to be directly synthesized using clean electricity to drive chemical reactions. This, Pillar 3, accounts for the remaining gaps of emissions reductions in the energy system that couldn't be solved with Pillar 2.

And then there is that third of global emissions which comes from non-energy processes. Pillar 4 is about shifting industrial processes and agricultural practices to emit less or not at all. This pillar encompasses a wide range of steps, from technology development to policy to outreach on farms around the world.

Finally, by 2050 significant levels of sequestration will be needed to put the negative-emissions "brakes" on global temperature rise. Whether or not that sequestration is possible—politically and physically—to pay for depends on technology development and scale-up for sequestration methods, and on having nearly eliminated global emissions. Sequestration, Pillar 5, is the bit that makes up for remaining emissions and gets the world to the goal of negative emissions.

In all, then, there are five pillars to solving climate change 100%:

1. Deploy clean electricity generation.
2. Electrify equipment that can be electrified.
3. Create synthesized fuels for equipment that can't be electrified or isn't electrified by 2050.
4. Implement various non-energy shifts.

5. Make up for the remaining emissions and get to negative emissions using sequestration.

Pillars 1–3 address 100% of energy system emissions, and Pillars 4–5 address 100% of non-energy emissions plus the need for negative emissions. Pillars 2 and 3, and much of Pillar 5, rely heavily on cheap, clean electricity being abundantly available from Pillar 1 to power all the new end-use, fuel synthesis, and sequestration equipment.

These five pillars are the physical transformations needed—at a minimum—to solve climate change 100%. No matter what combination of policies, initiatives, public actions, and private actions drive the enactment of these pillars, climate change impacts will continue to get exponentially worse until all five are fully implemented. To avoid the worst effects of climate change, that implementation must happen by 2050. So a 100% solution means implementing all five of these pillars in the next thirty years, which entails a huge transformation in energy, industrial, and agricultural systems, faster than any in history.

That is the scale of "what needs to be done." Now we need a plan that can add up to it. So far, we are wildly off track, still *increasing* the rate of emissions from year to year as almost all clean options fail to gain the initial momentum to scale globally. We don't have "all the technology we need" *affordable*

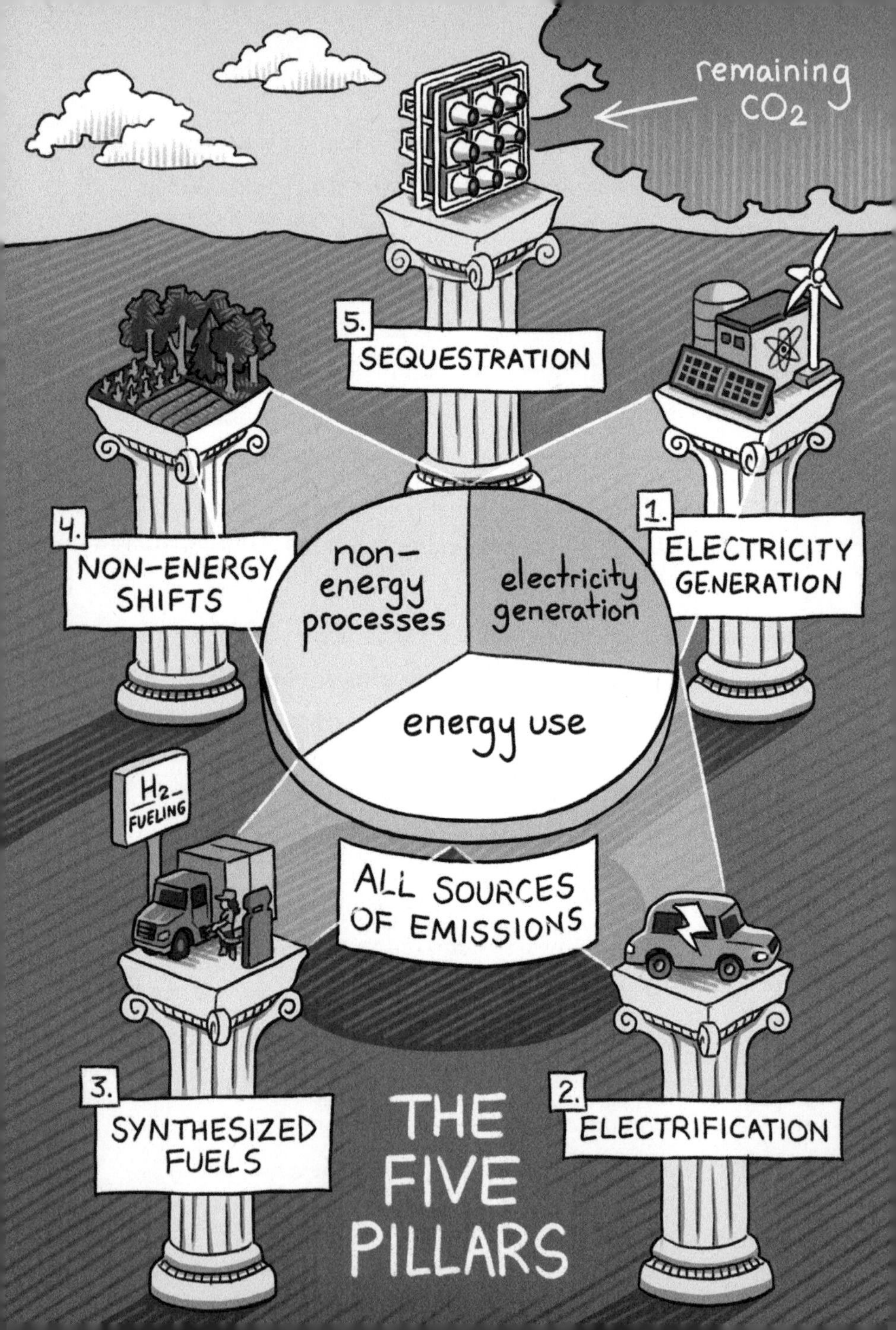
remaining CO_2
5. SEQUESTRATION
4. NON-ENERGY SHIFTS
1. ELECTRICITY GENERATION
non-energy processes
electricity generation
energy use
H_2 FUELING
ALL SOURCES OF EMISSIONS
3. SYNTHESIZED FUELS
2. ELECTRIFICATION
THE FIVE PILLARS

There are five pillars to solving climate change 100%:

1. **Deploy clean electricity generation.**
2. **Electrify equipment that can be electrified.**
3. **Create synthesized fuels for equipment that can't be electrified or isn't electrified by 2050.**
4. **Implement various non-energy shifts.**
5. **Make up for the remaining emissions and get to negative emissions using sequestration.**

enough even in countries where political will could mandate slight cost increases. There are large slices of the pie for which full decarbonization would currently mean massive increases in costs.

And then we face the fact that two-thirds of emissions come from developing regions, where air pollution makes clean options politically popular but where fossil fuels are still being expanded because they are by far the cheapest energy sources, pulling huge populations out of poverty.

To put the world on a path that gives us a good chance of reaching negative emissions by 2050, we need the clean options that make up the five pillars to be fully adopted to address 100% of the pie. For any given clean option, there are essentially two ways to make it happen:

- Make it affordable *enough* through some amount of technology/manufacturing improvement, and then add policy mandates or incentives in almost every country so it is adopted at full scale;
- Or, make it definitively cheaper than its emitting alternatives through technology/manufacturing improvement so that it outcompetes emitting options globally in the market economy. Generally, clean equipment comes with lower operating costs over time, so the challenge is bringing the up-front ("capital") costs down to levels similar to those of dirty equipment, which will lead to rapid adoption by companies and consumers.

The latter approach is generally faster, but where viable the former can give more certainty. The definitively cheaper approach requires massive tech improvement efforts in a short timeframe, while the modest-improvement-and-policy approach requires significantly more political will and campaigns across many different countries. Of course, some clean options will be better suited to one approach than the other. Both approaches require some level of innovation for most technologies, and both approaches may require incentive policies to drive innovation or ensure adoption of new technologies.

MOONSHOT PROJECTS

POLICY AND INNOVATION COMPLEMENT EACH OTHER

Many non-emitting options (technologies or new practices) could decarbonize a given slice of the emissions pie if rolled out globally. For each to do so, we must either improve the cost modestly and pass policy mandates or incentives in many countries, or improve the cost significantly so that the clean option outcompetes the emitting alternatives in the global economy without policy incentives.

This book uses "policy" to mean laws and regulations that require or encourage certain behavior. This includes, for example, mandates or subsidies. "Policy" here doesn't include actions taken by government bodies directly, such as deploying a set of charging stations or carrying out a project to coordinate fuel synthesis research. Most direct government "initiatives" that we'll discuss fall under the umbrella of work to promote "innovation," the development and improvement of technologies to lower their cost and increase their scale of deployment.

Some technologies are already "affordable *enough*" for policy to be feasible—for example, electric cars are cheaper to operate than gas cars but currently more expensive up front to buy. Subsidies or mandates could probably be made big enough to spur a rapid switch to electric cars. Some agricultural practices that reduce emissions can save money for farmers or improve yields at the same cost, and so they require only regulation of or education for farmers.

When we rely on policy to ensure that a given option rolls out fully by 2050, the relevant mandates or incentives must be adopted in most countries. If a few countries with insignificant emissions fail to adopt the policies in time, other countries would have to pay for sequestration to make up the difference, or the international community could use tariffs or sanctions to force cooperation from the remaining emitters.

Even convincing most but not all countries to adopt a given policy can be a slow and difficult road. International agreements and grassroots movements can shift what is politically viable and speed up the timeframe for adopting the relevant policies. Technology cost improvement can make mandates or subsidies more politically acceptable. But a 100% strategy that relies on policies still requires many individual political efforts—and that means lots of proactive decisions, which makes solutions less likely to scale fully in time, as noted in Chapter 1.

Innovation that makes clean options definitively cheaper, on the other hand, can spread solutions without the need for policy in each individual country. A small number of entities—as few as one if that one were the president of the United States or China—can start the required efforts to improve technology costs, and then sell those technologies to the rest of the world. However, there's no certainty that any given technology will improve in time to be definitively cheaper than its polluting alternatives. That's why policy mandate and incentive efforts should be pursued wherever they are viable—they may become necessary to close the gap to a 100% solution.

As noted before, "innovation" is used to mean both invention of new technologies or practices as well as the development of already-emerging technologies and practices to the point of having been proven, demonstrated at full scale, financed, and commercially deployed. Innovation also means improvements to existing technologies, including inventions of new components, but also advances in business or manufacturing processes that bring down costs, or public initiatives that bring technologies to larger-scale manufacturing and deployment to lower costs through pure economies of scale.

Innovation and policy complement each other. Policy in one country that ramps up deployment of a given technology will likely bring down the cost of that technology and thereby make it easier to deploy elsewhere. Innovation that directly decreases the costs of technologies or practices makes it more politically viable to mandate or subsidize those technologies or practices through policy.

However, there is one key difference between innovation and policy when it comes to achieving a 100% solution to climate change: policy can be repealed; innovation cannot. In Australia, the governing coalition passed a carbon tax in 2011. In short

> **Policy can be repealed; innovation cannot.**

order, a new coalition won power and repealed it. In the United States, the Obama administration created regulations to require more efficient cars and cleaner power plants. Within a few years, the Trump administration had repealed those mandates. Yet, over that whole period, coal power plants continued to be replaced by methane ones because methane had become the definitively cheaper fuel due to innovation on methane drilling techniques. Innovation-based solutions to climate change will likewise be more permanent than policy-based measures.

And because some technologies are currently too expensive or undeveloped for policy mandates to force their adoption, even in industrialized countries, some amount of innovation is needed by definition. Those of us hoping to see a 100% solution by 2050 should focus on making sure that innovation happens.

PRIVATE INNOVATION IS TOO SLOW

Activists and policy leaders in industrialized countries must focus on ensuring that every piece of necessary technology is developed and becomes affordable in time. We cannot afford to be distracted by "doing our part" through incremental reforms. Technology will take some time to ramp up to full-scale production and will take further time to be deployed everywhere, so the innovations to set each technology on a successful path must be achieved in the near future.

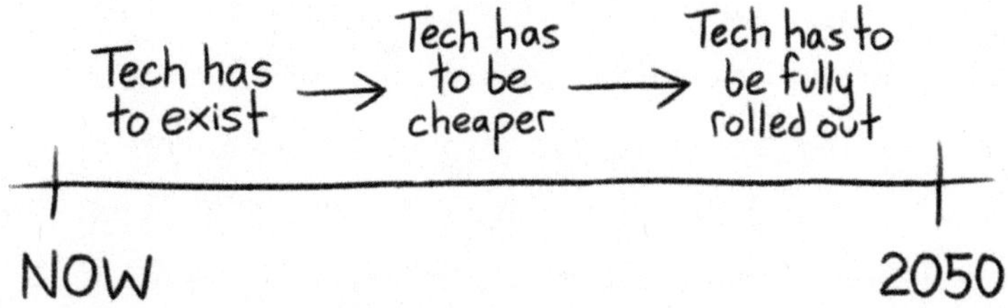

The exact timing varies depending on the technology, but this order of steps takes time and must be completed by 2050.

Private researchers at colleges and companies are constantly working on cleaner ways of doing all sorts of processes.

For example, a couple of startups recently started trying to commercialize a kind of seaweed that, when substituted for as little as 1–2% of cattle feed, virtually eliminates methane belches from livestock.[24] Companies such as Tesla and Chevy are already commercializing electric cars and starting to roll out electric trucks. Other startups are developing carbon-capture technology, while others still are devoted to the synthesis of carbon-neutral or carbon-free fuels.[25]

There's nothing physically impossible with the idea that this private innovation could solve the problem fully by 2050. But given the normal pace of innovation, the chance that researchers would hit upon, and companies would fully deploy, every single technology we need between now and 2050 is extremely low.

The slow pace of innovation has to do with the fact that innovations benefit society as a whole far more than they benefit their inventors. Economists call this a "positive externality" (the opposite of "negative externalities" such as carbon pollution—which hurts society as a whole much more than it hurts the companies or individuals that emit a given amount of pollution). Because of this fact, the economy systematically underinvests in innovation. That's partly why politicians of all stripes tend to agree that government funding should support innovation efforts: to bring benefits to society that individual people and companies wouldn't have enough profit-driven incentive to pursue.

Innovation funding tends to come from either government bodies—usually focused toward academic or very early-stage commercial research—and venture capitalists—usually geared toward proven, ready-to-scale technologies that need only a good business plan to take off. In between is a gap that observers call the "valley of death" for innovation. Many climate change solutions are currently floundering in this valley.

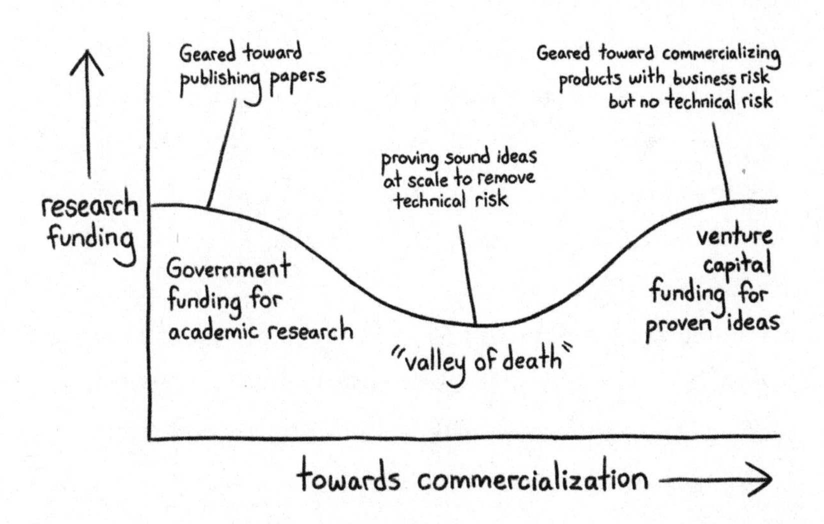

To dramatically increase the odds that we will find and fully deploy each needed technology, we should have public projects to spur the necessary innovation. This is essentially a larger-scale version of the existing consensus that government bodies should invest to correct the "positive externality" of innovation. In this case, we need those investments to be massive and focused on key realms of innovation that will bring the great benefit of solving climate change to society.

A MOONSHOT EFFORT

Let's focus for a moment on the United States, where this kind of public innovation has had storied success in the past. Military research has led to the commercialization of radar, GPS, the Internet, and more. Government-coordinated projects have mapped the human genome to accelerate medicinal inventions, developed both solar and nuclear power into viable electricity generation technologies, and put people on the moon.

Some people now call for a New Manhattan Project to drive innovation for climate change solutions. Others call for an Apollo Program. When Congress created an entity that oversees a degree of valley-of-death-stage innovation on clean energy, they named it ARPA-E after the (Defense) Advanced Research Projects Agency that helped make GPS and the Internet commercial realities. None of these are perfect comparisons for the type of government innovation we need today—the Manhattan Project simply required proving that already-understood science would work for a single invention and working out the engineering to build it. The Apollo Program is a better example: the US federal government funded and coordinated a massive set of efforts, from basic research to full-scale engineering and deployment. It partnered with companies in many ways. It cast a wide net in considering technologies, and focused on those technologies that turned out to have promise. It had a single goal, but needed many technologies to achieve it, and no one knew exactly which would be the successful ones at the beginning. However, the Apollo Program never had to scale technologies up to the point that they were commercially viable (if that were the case, we might have had commercial solar power decades sooner). And DARPA/ARPA-E are focused on applied research—more product-oriented than basic research funding, but still generally limited to lab-scale academic inquiry without the level of focus and coordination needed for climate change.

The youth-led Sunrise Movement and others call for a "World War II–scale" mobilization, which is perhaps the most apt comparison. During WWII, the US economy was dramatically transformed to supply the Allies with all the military equipment they needed. New technology was invented, such as practical radar. Existing technology—such as airplane manufacturing—was scaled up massively to bring costs down and provide the necessary amounts of equipment. Not only did it help win the war, but the massive industrial mobilization boosted the economy enormously.

FDR's leadership led to a dramatic scale-up of airplane manufacturing. In all, growing out of the Great Depression, the US economy grew 11–12% per year during WWII.[26]

It required ambitious goals, though. FDR's call early in the war to quadruple US airplane output to 50,000 per year was not unlike JFK's later call to put a man on the moon by the end of a decade. In both cases, these presidents were met with wild enthusiasm, but also skepticism about the likelihood of hitting such ambitious targets. In the end, the United States exceeded both goals. In 1944, the United States manufactured almost 100,000 airplanes;[27] by the end of 1969, NASA had landed not one but two Apollo crews on the moon.[28]

Similarly, the scale of deployment for climate change–solving technology will be massive. Many people will be skeptical about the idea that it could ever be possible to build enough new electricity generation to not only replace our entire fossil fuel grid but to power newly electrified processes, synthesize carbon-neutral fuels, and pull CO_2 from the air. And people may doubt that new cement production processes could ever roll out fully in a thirty-year timeframe, or that we could convince any significant percentage of farmers to manage their soil more sustainably. But if we could build those hundreds of thousands of airplanes in five years, and land those spacecraft on the moon in ten, we can surely transform the world's energy system in thirty if we innovate and scale up technology in a massive way.

The top goal of US climate activism and policy leadership should be to create moonshot-type projects to coordinate and fund innovation and to support the deployment of new technologies and new practices worldwide.

Activists and policymakers outside the United States may not have quite the same large economy and national budget to work with, but a coalition of smaller countries could carry out the same sorts of efforts. Individual countries can also tackle

The top goal of US climate activism and policy leadership should be to create moonshot-type projects to coordinate and fund innovation and to support the deployment of new technologies and new practices worldwide.

specific pieces with a mind to how they will fit into a 100% solution. (See Chapter 11 for more ideas on the role of various countries.)

COORDINATION AND FUNDING

Coordination is needed so that research and development efforts focus on technologies that *could* actually be a part of a 100% solution. Right now much research is based on what professors find academically interesting, what topics are likely to lead to new *scientific* knowledge that could be published in a paper, and what has long-established sources of funding. That will need to shift toward research focused around products that can be *commercialized* (and by 2050) rather than simply published in an academic paper. Through public projects, government bodies should direct willing partners to conduct research in areas that seem most promising or impactful and steer efforts away from avenues that have proven themselves to be too slow, impractical, or insignificant. Governments' ability to convene researchers and funders to support them can help highlight which avenues of work are most crucial. Public projects would also expand the scale of collaboration and knowledge sharing among innovators, thereby making it more likely the best minds will hit on the necessary ideas.

The normal pace of innovation might eventually get us to negative emissions, but not by 2050. As we will see in the coming chapters, the pace of innovation and deployment must accelerate by several orders of magnitude, which will require major funding increases. Presidential candidates in the United States have started talking about doubling the current worldwide level of clean tech innovation funding, but across basic research, testing, and early scale-up, the total will need to be far higher. For a rough order of magnitude, we should assume that effective moonshot projects would require a few hundred billion dollars per year of direct funding, which would be leveraged to shift something like 1–2% of global GDP (a couple trillion dollars per year) toward investments in clean tech deployment. This is in line with models that have projected global business-as-usual investments in fossil infrastructure vs investments of similar magnitude in clean infrastructure instead.[29]

Yes, that is a lot of money. But it is simply an up-front investment over a few decades to significantly decrease the costs the world would otherwise face, of building and maintaining fossil infrastructure ($2–3 trillion per year plus costs from climate change damage), deploying clean technology without having improved its cost through innovation ($4 trillion per year or more), or dealing with the impacts of climate change (probably also in the low trillions of dollars per year in addition to whatever infrastructure is paid for).[30]

Moreover, hundreds of billions of dollars per year is manageable within existing national budgets. The United States alone could dedicate that level of funding to moonshot-type projects on climate change, and the benefits of new innovations would boost the US economy in the end. On average, government-led innovation brings a 20% to 30% or higher return on investment to the US economy; some projects have brought much larger payoffs, such as the Human Genome Project, which is estimated at a 14,000% return on investment.[31]

Tens of those hundreds of billions could come from redirecting funds that are currently subsidizing fossil fuels. Some of the project's work could be done through the already-large military research departments, maintained as part of the $600-billion-per-year military budget. More could be done in partnership with private companies and labs, leveraging federal funds to coordinate many times more in total funding. And still more could be co-funded with other countries that want to share the economic benefits of such innovation.

STRUCTURE

The projects could be carried out by a new agency or agencies created for this task, or could be a set of initiatives among existing agencies. Some new entity might be preferable, because it could be more independent. Currently, Congress funds many important initiatives, but wastes money by requiring various projects to happen in certain legislators' districts. Sometimes legislators pressure federal agencies to do the same, and sometimes they succeed, but there is good precedent for agency leaders resisting such pressure. (Energy Secretary Steven Chu, for example, was known among members of Congress for politely listening to such requests, then reiterating that his agency made decisions based only on maximizing efficiency and impact.[32])

A new quasi-public, independent agency funded in large lump sums from Congress (or the president, who could redirect a substantial amount of money toward such projects even without Congressional approval) would minimize such politics-based decisions. It could then coordinate with and fund other agencies and private partners, aided by the White House when there is a supportive president who can add extra clout to its work.

This model could also apply to non-US-led projects: a coalition of European and Asian countries, for instance, perhaps with the partnership of some international companies, could

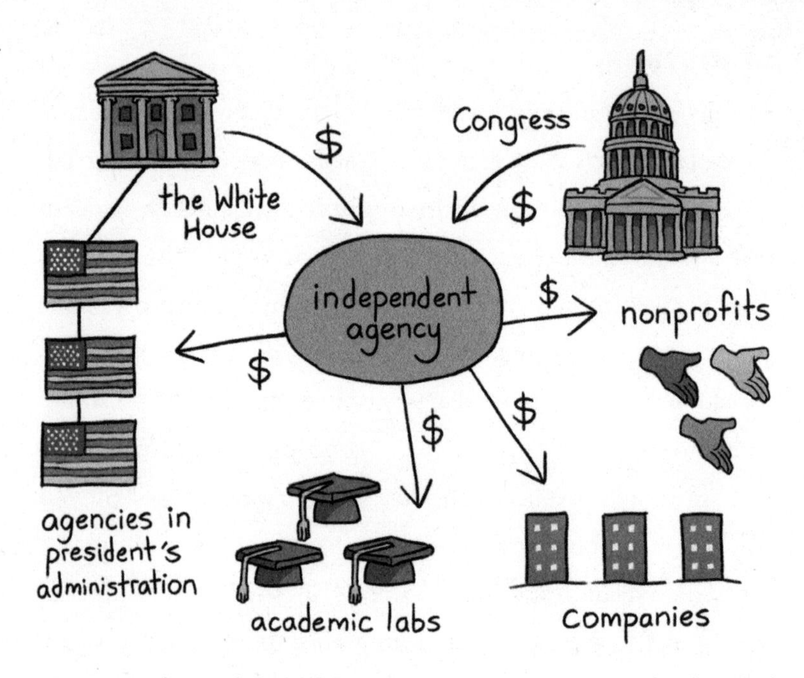

give an independent NGO or new UN agency a similar level of funding. Such an organization could then coordinate and fund various entities and efforts around the world.

METHODS

Some technologies will require basic research, which the relevant agency or agencies can fund and coordinate. Some will require testing and demonstration, which could benefit from centralized, publicly available labs and test beds built or co-funded by government investments. Some will require help with initial scale-up—a tricky phase of commercialization in terms of our climate change–solving timeframe.

Scale-up brings the cost down from the "first of a kind" unit to the "nth of a kind" unit, which is usually much cheaper because of having an established and honed manufacturing process, steady supply chain, expertise learned by workers, and economies of scale.

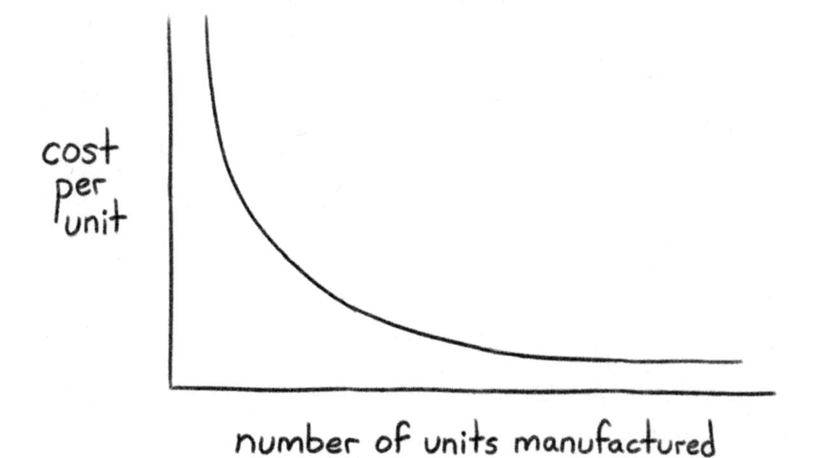

Project leaders will have to find ways to coordinate early adopters of new technologies, including in some cases federal government agencies, and sometimes help fund the production of the first batch of units of a new technology to reduce the time it takes to fully establish a manufacturing process and bring costs to competitive levels. For non-energy shifts, efforts and funding support will have to focus on direct initiatives around the world, for instance to educate farmers about the benefits of better soil and fertilizer practices.

Much of these projects' work will be proactive—identifying specific technologies needed, convening industries and researchers, reaching out to international partners—but much will also be reactive. Agencies will have to continually assess progress and prospects for various technologies, note when any have proven impractical and adjust accordingly, identify new options worth exploring, and decide in each moment the best distribution of a limited amount of money. Unlike a policy, for which language is written ahead of time and implemented once a bill is passed, the projects will start with only a vision of what they need to develop and a framework of how they will work. They will have to implement all these ideas with real people—farmers, company leaders, agency administrators—as partners, and react nimbly to changing technological or economic situations.

THE CORE OF THE 100% SOLUTION

Such projects will bring us as close as possible to guaranteeing that the world will achieve negative emissions by 2050. Conveniently, innovation is popular and permanent.[33] And the scale of funding required, though large, is not unreasonable for a small number of countries to cover. Through a focus on innovation, a tiny number of global leaders could successfully set the world on track toward a 100% solution.

The five pillars form the blueprint for exactly what these coordinated innovation efforts must achieve. The pillars are numbered in roughly the chronological order that their emissions reductions must play out, starting with electricity generation and ending with sequestration. The following five chapters explain the specifics of each pillar, the current state of innovation and relevant policies that could be pursued, and the most crucial steps for public projects to take to ensure that clean options are developed and adopted in time.

PART 2

WHAT HAS TO HAPPEN

5

PILLAR 1: ELECTRICITY GENERATION

GROWING DEMAND

By 2050, billions of people will have new or increased access to energy. If clean energy becomes cheaper than expected—as it must in many cases to spread rapidly enough in developing countries—even more people will be able to afford larger amounts of energy. And the amount of sequestration required by 2050 will demand significant additional energy inputs.

Nearly all of this growing demand for energy services will

have to be met with electricity—either directly, in the case of equipment that is already or will be electrified; or indirectly, through synthesized fuels (and some amount of biofuels) that can substitute for fossil fuels and which require large amounts of electricity to synthesize. Currently the world generates slightly under 25 petawatt-hours (PWh) of electricity per year. Most models show this doubling to around 45–50 PWh per year by 2050.[1]

However, if we want to guarantee negative emissions by 2050, we have to aim for a dramatically higher number. Models that project around 45–50 PWh of electricity generation per year either show continued large amounts of emissions, with only modest increases in electrification, or they rely on the assumption that nearly every country will impose stringent efficiency policies and that energy demand will decrease dramatically even while economies continue to grow. Models that show near-zero emissions by 2050 also all rely on significant amounts of biomass-based fuel substituting for fossil fuels. As we'll discuss in Chapters 7 and 9, this can certainly happen to some degree, but a strategy that relies so heavily on biomass would require strict regulation in every country (which we'd have to count on not only being implemented, but not ever being loosened once the energy system was dependent on so much biomass), which is highly unlikely in our thirty-year timeframe.

Under a framework that charts a path to negative emissions by 2050 *without* relying on strong regulations in every country, we must expect that electrified options (direct equipment and synthesized fuels) will have to be made significantly cheaper, and possibly subsidized by solution-focused countries, so they spread rapidly across the world. We must assume that fossil fuel use with carbon capture and storage ("CCS," which most models also rely heavily on) will be limited to countries with strong mandates or incentives (see later in this chapter). We must assume that only a modest amount of biomass can contribute to the total energy demand, given the likelihood that growing biomass for fuel would contribute to deforestation emissions absent global regulation. We must assume that some amount of direct sequestration of atmospheric CO_2 is needed to close the gap to zero emissions and achieve negative emissions.

And therefore, we must assume that the world will need not 45–50 to fifty PWh of electricity by 2050, but more like 100–150 PWh. This is about the range we'll need if 2050 end-use energy demand meets the current-trajectory projection but we electrify roughly 60% of energy use (the upper bound of most electrification models,[2]) with some portion of the other 40% served by synthesized fuels (the rest served by fossil fuels with CCS and biofuels), and with levels of electricity-powered sequestration in line with those discussed in Chapter 9. If countries don't come through with efficiency measures to meet their existing energy commitments (e.g., those from the Paris Agreement, which

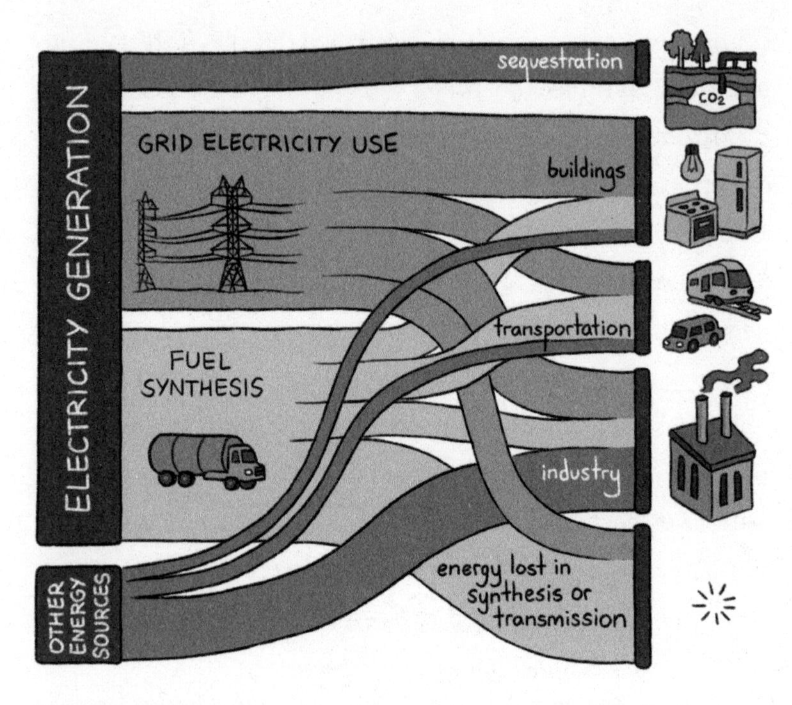

The energy system in 2050 must be mostly powered by electricity—either directly through the grid, or indirectly through the electricity-powered synthesis of clean fuels. Electricity will also be needed to power some portion of the required sequestration each year. Biofuels and geothermal, and, in countries with strong policy, also fossil fuels with CCS, will make up the remainder of total energy demand.[3]

indeed no major country is on track to meet), total demand and the need for non-regulatory solutions could push this number even higher, toward two hundred PWh per year of electricity generation. Thus, depending on how much policy actually gets implemented, electricity generation might need to grow to anywhere between two and eight times its current levels by 2050—most likely about four to six times.

Most activists acknowledge that we are not on track to replace the current-size electricity generation system with clean equipment. Most models project that we will in fact need to build a clean electricity generation system double the current system's size. But in reality, we have to plan for a solution that doesn't rely on massive efficiency measures in countries like India, nor on strictly enforced biomass-growing regulations worldwide. A 100% solution can still be readily achieved without every country's long-term cooperation, but only with two to three times more clean electricity generation than anyone is talking about.

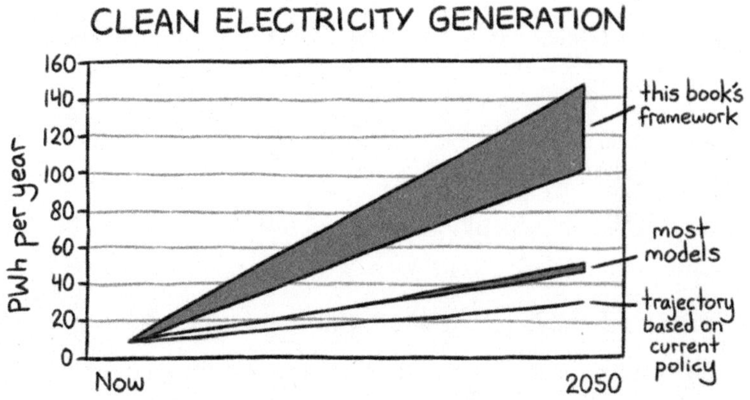

If the goal is certainty under all policy scenarios for a negative-emissions 2050, we are two to three times more off track than most people are thinking.[4]

There's one more challenge: most solar and wind facilities are only rated for thirty years or less, and though nuclear facilities last somewhat longer (forty to sixty years or sometimes more), most of those currently in use were built quite a while ago.[5] So, most of the clean electricity generation capacity currently installed will be retired by 2050 (some hydro and bits of

A 100% solution can still be readily achieved without every country's long-term cooperation, but only with two to three times more clean electricity generation than anyone is talking about.

the others will remain). This means that the electricity generation system we need—one four to six times larger than what we currently have—will have to be built in thirty years almost from scratch.

RAPID BUILDOUT

That's already a difficult undertaking, but the world should aim for the even higher goal of deploying as much of that new generation as possible in the *near* future rather than waiting until the 2040s for most of the electricity generation transition. Partly that's because the sooner any given sources of emissions are eliminated, the less cumulative CO_2 will be in the atmosphere by 2050 and therefore the less sequestration will be needed. Because of its centralized nature, fossil fuel–based electricity generation, especially from coal power plants, holds the single biggest chunk of emissions that can be eliminated in the near future.

Furthermore, the sooner clean electricity generation is definitively cheaper than fossil electricity generation, the easier it will be to deploy, because it can compete with *new* fossil equipment that would otherwise be built. Convincing a company or country to build a solar or nuclear plant instead of a new coal plant they would otherwise build is fairly easy if the clean option is cheaper. But convincing them to replace a just-built, operating coal plant with a clean plant is very hard. The payback would have to be extremely fast—maybe clean electricity generation can get there eventually, but to maximize the chance of fully eliminating fossil fuel electricity generation by 2050 or sooner, it would be wise to frontload as much of the deployment of clean generation as possible so that it can mostly compete with potential new fossil plants, not existing ones.

That's a large obstacle given that a key limitation of all clean electricity generation sources is how expensive they are up front despite costing little or nothing to operate (in the case of wind and solar, most important is the up-front cost of storage and load-balancing technologies to enable significant use of these intermittent sources). At least one clean generation option must come down in *capital* costs, so that when short-term decisions are made based on the cheapest way to add an additional bit of electricity generation output, a clean option is always chosen over fossil fuels.

SOLAR

If our window for solving climate change were more like a hundred years, solar might be the perfect technology. Millennial author Varun Sivaram laid out all the potentials and limitations of solar in his insightful book *Taming the Sun*.[6] Solar comes with great benefits for the long-term: it's easy to scale from tiny to huge installations, so you can test out new solar cell technologies without investing huge amounts of money. Eventual technologies for organic or "perovskite" solar cells will be printed on roll-to-roll manufacturing machines, a form of rapid, continuous production that will be much faster and cheaper than current batch-based and energy-intensive silicon processing.

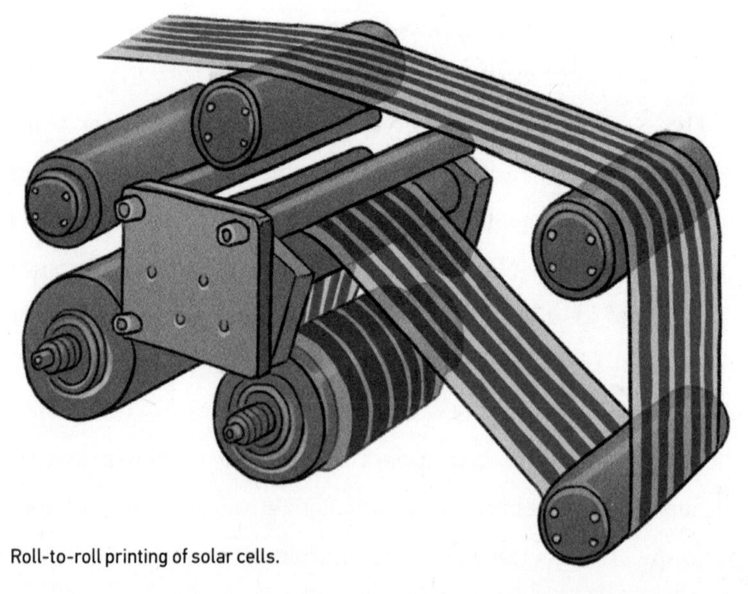

Roll-to-roll printing of solar cells.

While such roll-to-roll solar cells may eventually become extremely cheap, there are two problems. First, they aren't there yet, and we need something ready to scale up for the initial short-term electricity buildout the world requires (they might still contribute meaningfully toward the total system by 2050). Second, to power a high percentage of the electricity generation system, they'd need batteries (or other forms of storage or grid flexibility), which are subject to the same development and manufacturing ramp-up constraints with less certainty for success in low cost. That means that although solar might contribute heavily to the 2050 system, it can't form the majority of electricity generation by then. Eventually, by the end of the century we expect battery technologies will have caught up and solar will be viable to use for a larger portion of generation.

And so, solar should be expanded massively, but given the enormous amounts of material and land required for solar manufacturing and deployment, and the lack of cheap batteries, it will probably amount to less than a third of the total 100–150 PWh demand for electricity generation by 2050. Indeed, Sivaram, who is optimistic about solar's long-term potential and who works on commercial-scale solar deployment, puts meeting one-third of global demand as the upper goal for 2050 under the normal-assumption scenario requiring only 45–50 PWh of total generation.[7]

To get there, research and testing will have to pursue

organic and perovskite roll-to-roll printed solar cells, so that the cost of solar keeps declining in the next few decades. The more solar is deployed on the grid, the more costly it becomes to deploy because of the need to keep flexible fossil fuel generation, or storage, available to run at a moment's notice (but not to use most of the time, which makes that backup uneconomical in itself). After a point, additional solar will generate electricity almost entirely during times when excess electricity is already being generated by existing solar, making the value of new solar drop as more is deployed. To continue being economical, its costs will also have to drop, and energy storage innovations, as well as policy designs to transform grids, will have to be deployed. These could include adding more long-distance lines to bring excess solar from midafternoon in Western Europe to serve evening demand in Eastern Europe, or adding more responsive end-use equipment (for example, heaters that choose the time of day to heat based on electricity availability).

In the near term, policy mandates can keep accelerating solar deployment and can help bring new solar technologies to the demonstration and early commercialization phases. Policymaking and activist messaging should keep in mind that the physical realities of engineering a grid with solar are seriously different in different parts of the world. In northern regions, solar produces so little compared to its theoretical peak output that it can only contribute to a form of "efficiency," that is, reducing total demand for other generation but not shifting the source of that generation away from fossil fuels. In regions with strong winters, the difference in solar production in summer vs winter is enough to require a backup generation system (or months-long storage system, which currently no technology can come close to providing at any reasonable cost) to replace most of the solar generation during the winter. In regions closer to the equator with more consistent sunshine, solar faces few of those problems and merely requires hydro or methane backup (or nuclear backup, or storage, if either is developed to be affordable soon) for nighttime.

WIND

Wind provides the cheapest renewable electricity generation in many regions that don't have abundant hydropower.[8] And unlike hydro, which requires specific conditions that are not abundant worldwide, wind can work in a larger number of places. The total potential for wind generation is not nearly as much as the *theoretical* potential for solar, but it is far more than hydro and possibly close to the *economical* potential for solar. Wind could probably be scaled up to power about a third of total demand by 2050, if it was pushed significantly.[9]

Unlike solar, though, wind has already been improved technologically to achieve about the lowest costs we can expect.[10] Offshore wind dramatically expands the areas viable for wind deployment, but is significantly more expensive than onshore wind. Like solar, onshore wind can compete on cost with methane power plants *when the wind is blowing*, but wind also suffers from the problem of intermittency, and from the large amount of land and materials required to harvest a diffuse energy source. All the same grid and demand policies and storage technology improvements that must be applied to solar could also make wind a viable provider of a larger share of global electricity generation demand.

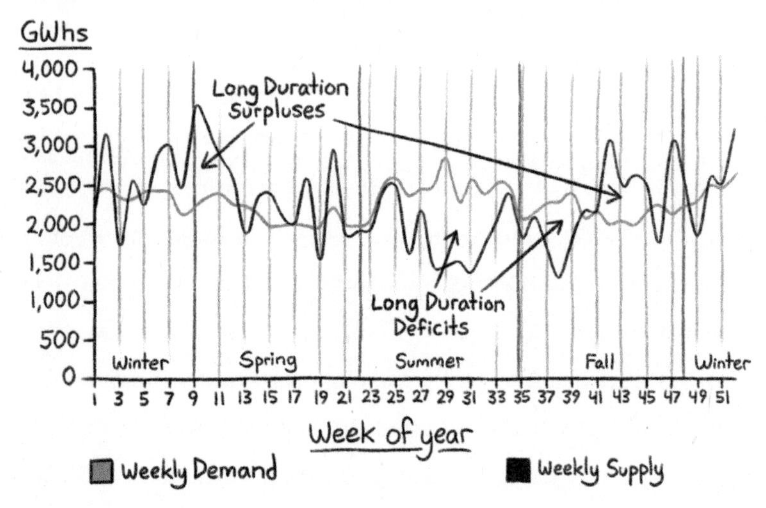

In this analysis from energy consultancy NorthBridge, which modeled a hypothetical 100% wind and solar system for New England (with the proportion of wind and solar that minimized surpluses and deficits), electricity would have to be stored for many weeks to several months at a time.[11]

Together—depending on how dramatically solar and storage technologies improve in cost, and on the volume of flexible grid demand that is built—solar and wind might be able to power one- to two-thirds of the total 2050 electricity generation demand. Much of the buildout, however, will likely happen in the second half of the transition, partly because it will take time and innovation effort for solar and storage technologies to drop further in cost, and particularly because the flexible demand loads that will enable much larger percentages of solar and wind to be economical on the grid will be much more common when fuel synthesis and sequestration scale up. To the extent that solar and wind buildout can accelerate, policy incentives and mandates will have to drive most of the increased pace for the next decade or two.

HYDRO

Hydro is currently the largest clean electricity generation source in the world, but the limited availability of good sites for hydroelectric dams restrict where it can be deployed, and most of the best ones have been taken. As discussed in Chapter 2, hydro is cheaper than fossil fuels where it is plentiful. Because it is a fairly centralized and concentrated form of energy, one new hydro project adds a lot of generation, meaning hydro can grow quickly as an electricity source. But new sites will encroach on natural ecosystems and native lands, and will therefore face political hurdles that limit the pace of buildout. Small-scale hydro faces fewer hurdles, but also has a much smaller potential for adding up to large amounts of generation.

In total, hydro may be able to double globally, possibly triple if we pushed it to the extreme.[12] Even a tripling would still cover only about 10% of the 2050 electricity generation demand.

The advantage of hydro for the larger electricity system is that it is "dispatchable," meaning it can generate more or less in a given instant depending on grid demand. Water gets stored up behind dams and released at peak demand times. The rate of water flow can be varied to alter the power moment to moment. This means hydro can serve as an excellent backup to solar and wind, filling in for their generation when the sun goes behind a cloud or sets for the night and when the wind calms. The more hydro that is added, the more solar and wind will be viable to add in the same regions.

NUCLEAR

Unlike solar and wind, nuclear deployment is not currently accelerating. In fact, for several decades the nuclear industry has totally floundered, and in Europe and the United States the costs for the few new nuclear plants that have been built have been extraordinarily high. However, in China and South Korea nuclear has been built in the past decade for about the same costs as some fossil fuel plants (in Europe and the United

States in the early days of nuclear plant deployment, it was even cheaper).[13] There remains some hope that if scaled up with better supply chain and manufacturing practices, nuclear costs could drop below those of the cheapest fossil fuel plants.

If this can happen, the advantages are tremendous. Nuclear is the most concentrated source of energy we know of for generating electricity. That means that the amounts of land, raw materials, manufacturing labor, and waste are all much smaller than those for solar or wind. It also means nuclear can be built out faster in terms of generation capacity.[14] And unlike solar and wind, nuclear power is not intermittent, and tends to run near its peak capacity almost all the time. That means that for every unit of peak capacity added, nuclear is generating far more actual electricity than solar, wind, or hydro.

So nuclear holds the potential to add clean electricity generation at the incredibly rapid pace needed. There are two problems: politics and economics.

The political problem is that people are scared of nuclear power, especially nuclear waste. This may be starting to change among the most intensely focused Millennial climate activists—for example, the Sunrise Movement has usually been careful with its rhetoric to refer to "clean" rather than "renewable" electricity so as to leave the door open to nuclear ("renewable" means that the supply is theoretically inexhaustible as long as the sun is shining on the earth; as nuclear permanently consumes small amounts of fuel it is not renewable, but is nearly inexhaustible for practical purposes)—but generally these fears are still pervasive. The idea of radiation triggers disgust and taints nuclear power as something "unnatural." Scientific measurements point in a different direction: even during the famous Fukushima nuclear incident, radiation exposure outside the plant never exceeded medically established limits, which are themselves well below natural levels of background radiation that exist in some locations in the world.[15] We're all walking around in a soup of radiation, but we don't usually think about it.

Radiation from nuclear power plants has only ever killed people—or even made anyone measurably sick—once, at the Chernobyl power plant. At that plant, which did not have the containment vessel that all reactors are built with today (and always have been, outside the Soviet Union), operators decided to carry out a foolhardy experiment with few safety precautions and a perfect storm of conflicting personalities. As the experiment started to go wrong, a series of bad decisions led to a set of explosions and the destruction of the reactor. Two people were killed by the explosions, and a few dozen more from acute radiation exposure. From all the radiation released, estimates put the upper limit of potential deaths over the next few decades at about four thousand (more recently, even that number has been called into question and numbers lower than

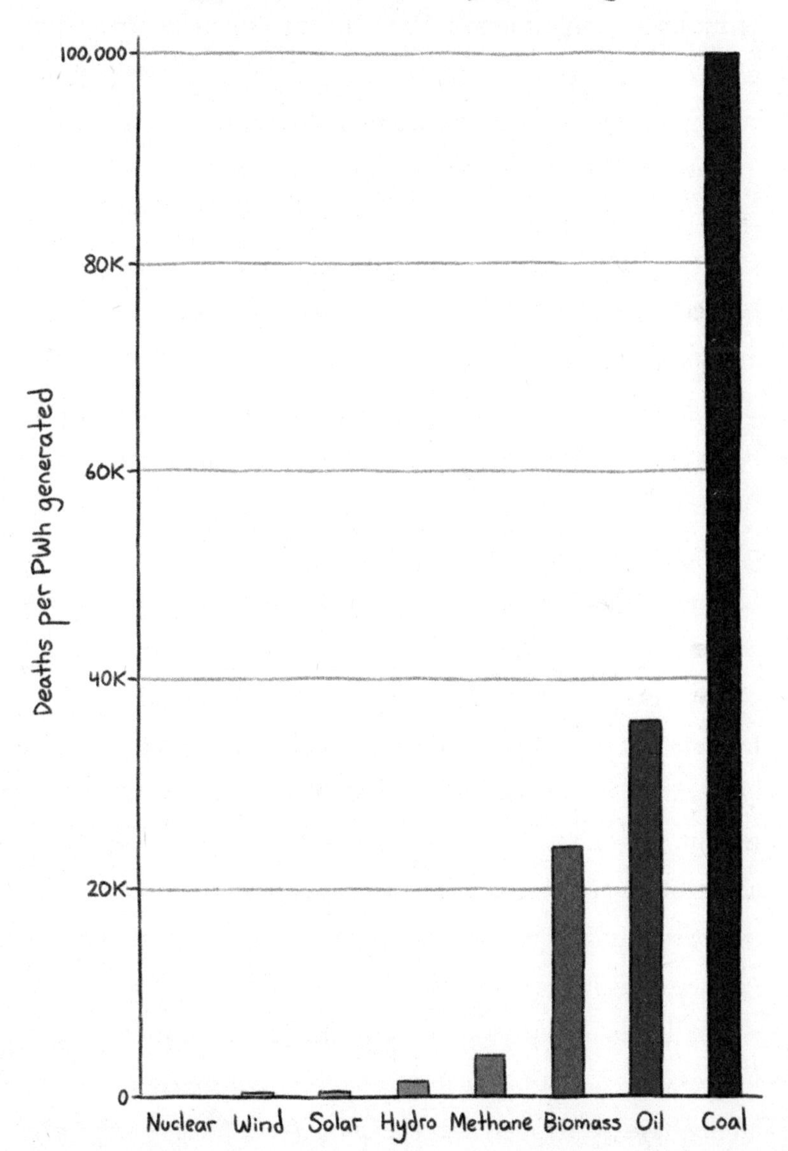
ELECTRICITY DEATHS
100,000
80K
60K
40K
20K
0
Deaths per PWh generated
Nuclear
Wind
Solar
Hydro
Methane
Biomass
Oil
Coal

one hundred have been suggested).[16] For perspective, four thousand is about the number of people killed by coal every couple days—mostly from respiratory illness caused by the particulate matter released when coal is burned.[17]

Nuclear is in fact the safest form of energy we have—even slightly safer than solar (which sometimes kills installers who fall off roofs) and much safer than hydro (which occasionally kills whole communities when a large dam breaks and towns are flooded).[18]

The revulsion people have about nuclear waste is similarly connected to a philosophical uneasiness about radiation. But this is perhaps the most misplaced concern of all those held against nuclear: the waste involved is easy to contain, so radiation from nuclear waste has never been a problem. And because nuclear is such a concentrated form of energy, its waste comes in far smaller volumes than waste from fossil fuels. In fact, nuclear produces even less waste than solar because of the chemicals used in manufacturing solar panels and the shorter lifetime of solar cells themselves.[19] All the wastes in question are toxic, but in this case radiation is actually an advantage: over time, radioactive waste decomposes and reduces its own volume. The other wastes stay toxic forever.

For true perspective, we must consider the question of wastes in the context of climate change: we need massive, immediate additions of clean electricity generation. We don't have long-term sustainable disposal systems for nuclear, solar, or coal wastes. We're currently generating massive amounts of coal waste, and to replace it we're going to have to generate more (but far less massive than coal) solar and nuclear wastes. Those, especially in the tiny volumes that nuclear waste comes in, can be safely stored in "intermediate" storage for many decades, giving us time to focus on the dramatically more urgent problem of solving climate change. Eventually, we can re-focus on developing sustainable disposal systems for all of these wastes—maybe by that time, current proposals for nuclear reactors that can consume (and therefore dispose of) nuclear wastes as fuel will have even been developed to commercial viability.

Roughly-proportional amounts of waste from coal, solar, and nuclear visualized.[20]

The immediate *economic* problem of nuclear power is much more pressing. Even the cheapest nuclear plants built in the last decade are too expensive to outcompete fossil fuels globally, let alone do so in the immediate future. To bring the costs of nuclear down enough to rapidly outcompete coal and methane, the key is standardization. So far, nuclear reactor designs have changed every few reactors built. Any single design has been executed a couple dozen times at the most. This means the technology is constantly starting at the beginning of the cost-lowering scale up process that solar and lithium-ion batteries have undergone dramatically in the past two decades. "First of a kind" plants are always much more expensive than "nth of a kind" plants.

Bringing down the cost of nuclear through standardization and scale-up of manufacturing is also the most likely route to the short-term addition of clean electricity generation we need.

Government efforts can be directed to convene industry players and spur adoption of best practices—for example, construction planning practices to minimize financing costs (loans taken out to pay for each new plant). Public funding, through direct procurement, power purchase agreements, or simply loan guarantees, can kick-start the construction of the first batch of standardized plants.

South Korea and China both have existing designs that they have built several times at reasonable costs.[21] International collaborations to begin mass-manufacturing those designs could be the fastest way to add clean electricity generation immediately. Such collaborations, though, might be hindered by the recent shift in South Korean political inclinations against nuclear, akin to stigmas in the United States and Europe. And Chinese politics are always hard to predict because of China's centralized government system. Nonetheless, leaders who could collaborate with them should explore a scale-up in exporting one or both countries' reactors.

At the same time, there are many designs for advanced, "fourth generation" nuclear reactors, and a few dozen startups trying to commercialize their version.[22] None of these are on track to add large amounts of clean electricity generation as rapidly as we need, but public convening, funding, and support (including testing and demonstration facilities, and guaranteed government purchases if new designs meet certain targets) could accelerate these fourth-gen reactors to commercialization in time to contribute to the short-term buildout of clean generation we need. Many fourth-gen designs have other benefits for the eventual electricity system, such as flexibility to ramp up and down at a moment's notice (a feature known as "dispatchable generation"), fewer operators needed, passive safety features to ease public concerns and make the politics of deployment easier, and modular sizes to allow deployment in a wider range of locations.

Often, proposals for cheaper, standardized nuclear plants revolve around mass-manufacturing them in a factory or shipyard. Shipyards are known for on-time and at-cost delivery of durable large industrial products. There are only a few with enough size and quality to contribute to a rapid buildout of nuclear, so more would have to be built soon, but they could each process dozens of plants per year. These plants could be

With smart business management, mass-manufactured nuclear plants could probably drop below the cost of the cheapest fossil fuel plants in the near future.

floated to their eventual deployment sites (some proposals even suggest skipping the step of towing them onto land and preparing a construction site and instead operating the plants in the ocean or a river right next to shore).

With smart business management, mass-manufactured nuclear plants could probably drop below the cost of the cheapest fossil fuel plants in the near future. Scaling up the supply chain (uranium mining and enrichment, shipyards and their workforce, turbine manufacturing, and the manufacturing of the nuclear plant) fast enough to transition significantly from fossil fuels to nuclear in a decade or so is a tall order, but within the realm of possibility.

FOSSIL FUELS WITH CCS

The final option for decarbonized electricity generation is to continue using coal and methane, but to filter out the CO_2 from the plant exhaust (some processes actually separate the CO_2 before burning the fossil fuel, leaving pure hydrogen to be burned and power the turbine), transport it to appropriate geologic sites, and bury it deep underground where it will not escape for a long time, if ever.

It might be a stretch to call this carbon capture and storage (CCS) *clean*, in that coal causes some degree of pollution, disease, and death (especially among coal miners) even before it gets burned, and methane causes large amounts of greenhouse gas emissions when it leaks during extraction and distribution (see Chapter 8). But fossil fuel power plants with CCS are certainly much *cleaner* from a climate change perspective. If political leaders can accept the continued health damage from coal (and hey, they've accepted it for a century already), then coal with CCS would be essentially emissions-free. Methane with CCS could be considered emissions-free if and when methane drilling and pipeline technologies improve to prevent leaks. Because of these uncertainties, CCS isn't explicitly a part of this five pillar framework—it could of course decarbonize factories without electrifying them or using synthesized fuels, but we don't know at the moment whether the methane ones would be truly zero emissions, or whether CCS in general has a significant chance of scaling.

The latter uncertainty is because of the economics of CCS: by definition, CCS involves extra equipment and extra steps on top of any industrial process. It will always be cheaper to run the exact same process without CCS. Therefore, virtually no company will ever adopt CCS technology for economic reasons alone. It will happen only in countries that either mandate it or put a significant enough price on carbon pollution that it becomes cheaper for a factory to use CCS technology than to pay the carbon price for its emissions. Part of the reason CCS doesn't feature in this framework is this fact that policy is absolutely essential for its widespread use, which makes it unlikely to decarbonize as large a portion of electricity generation and industrial emissions as technologies that can come down in cost to the point that they actually save money.[23]

However, all that said, CCS has one major advantage in countries where it is politically viable to mandate its use: CCS equipment could be added to most existing power plants for some significant but not prohibitive cost, and that would mean an extremely rapid first step in reducing power plant and factory emissions. In some countries, CCS might play a significant role in the 2050 energy mix, either because of political choices to allow fossil industries to transition less abruptly and therefore win support for larger policy proposals, or as a last-mile effort to eliminate remaining emissions when we approach 2050.

In other countries, CCS might not play a major role by 2050, but it does present a major opportunity for immediate emissions reductions in countries willing to enact sweeping policy. Remember that the sooner large sources of emissions are eliminated, the smaller cumulative amount of emissions will need to be sequestered from the atmosphere in 2050 (and atmospheric capture and sequestration is much more expensive

CCS equipment could be added to most existing power plants for some significant but not prohibitive cost, and that would mean an extremely rapid first step in reducing power plant and factory emissions.

than factory or power plant based capture and sequestration). CCS is highlighted here as a possible intermediary measure in countries with the political clout to use it, while we start mass-manufacturing standardized nuclear plants, and while solar and storage research is pushed forward to take a larger role later in the transition.

OTHER EXISTING TECHNOLOGIES

Aside from solar, wind, hydro, nuclear, and fossil fuels with CCS, a couple of existing technologies could supply modest portions of the 2050 electricity generation mix. Geothermal heat can power some industrial processes, heat buildings and water, and sometimes create steam to generate electricity with.[24] Existing geothermal technology is limited to locations with the right geologic features and doesn't have massive total potential for the energy that could be harvested globally, so it will likely play a small role.

Biofuels can substitute for fossil fuels, though as we will discuss in Chapters 7 and 9, they can only be scaled up modestly before they start causing more emissions, through deforestation, than they save by replacing fossil fuels. There are various kinds of biofuels, and some (such as from anaerobic digesters on farms) may be more sustainable than others (such as ethanol from corn that takes away land from food crops). Some models of a decarbonized 2050 rely on vast amounts of biomass energy,[25] which could be viable if most national governments impose ambitious regulations. But we cannot count on them to be carbon neutral in the more likely scenario in which national governments don't implement strict enough rules or monitoring programs to guarantee biomass is grown without deforestation impacts. Therefore, biomass figures minimally in this framework, which is meant to show the viability of solving climate change *without* the ambitious and sustained international cooperation that currently seems unlikely.

EMERGING TECHNOLOGIES

Moonshot efforts for electricity generation should focus most immediately on a rapid scale-up of nuclear (or CCS where viable to mandate) to steeply reduce electricity generation emissions. These projects should also aim to bolster the acceleration of solar and wind deployment, particularly with a mind toward the extra level of competitiveness solar (and storage) might reach for the second half of the needed energy transition when more varied uses for electricity will become prominent.

At the same time, moonshot efforts should immediately start supporting basic research or "valley of death" phase work on new electricity generation technologies that may have no certainty of contributing, but that could turn out to play a major role if they were developed successfully.

Perhaps the most promising proposal is for deep geothermal heating and electricity generation, which can be deployed in far more locations than traditional geothermal, and which could supply higher temperatures to run power plants and some factories with no intermittency.[26]

Another idea is for generating thorium-fueled (rather than uranium-fueled) nuclear power.[27] One advantage is that key large countries, including India, have more abundant domestic supplies of thorium than uranium.

Other ideas include undersea water current power, which is akin to offshore wind turbines, but has the turbines spun by water near the bottom of shallow ocean areas. Related proposals include tidal power and wave power systems.[28]

What underwater current turbines might look like.

Nuclear fusion (rather than fission, which all current nuclear power is) has been said to be "thirty years away" for the last fifty years, but it would indeed be revolutionary if we could commercialize it successfully.[29] It deserves some level of research and testing support.

And then there are more out-there ideas that may be in the earliest lab stages. Who knows what someone may discover and whether it could have a chance of scaling to commercial usefulness? A portion of funding should be used to help along these innovations with low probabilities of success but with serious game-changing potential if they succeed.

STORAGE TECHNOLOGIES

Finally, solar and wind can only reach a high percentage of electricity generation if electricity storage technologies become much cheaper. Nuclear would also benefit significantly from the deployment of storage: Unlike solar and wind's uncontrollable fluctuation with weather, current nuclear plants are hard to turn on and off, so they run at a steady power almost all the time, even when grid demand fluctuates. As nuclear becomes a large share of electricity generation, there may be times when the total output from nuclear plants is considerably more than total demand at that moment. Affordable storage technologies would improve the economics of this grid system tremendously.

The storage in question could be lithium-ion batteries, as dominate now. They will probably come down somewhat more in cost. But current projections don't show them coming down enough that weeks or months of storage could be viable.[30] Other batteries may be invented or scaled up—already, "liquid flow" batteries are being demonstrated in early commercial deployment and show promise of eventually beating lithium-ion costs for grid-scale storage.[31]

The cheapest current form of electricity storage is pumped hydro—the practice of pumping water up a hill when there's excess electricity on the grid, and letting it down through a hydro turbine when electricity is needed. This requires specific sites (mountains next to rivers or lakes), so its potential is limited, but it could be scaled up a bit. Similar gravity-based storage systems have been proposed, such as ones using concrete blocks and cranes, though none have shown serious commercial promise.[32] In the physical storage category, pressurized air systems can store electricity, but they require either specific geological formations to pump air into, or undersea bags of air that get inflated.[33]

A different and more likely category of storage is using excess electricity to synthesize fuels—some of which can simply be the synthesized fuels sold to other sectors, and some of which could power traditional power plants (or fuel cells) and therefore act like giant batteries.

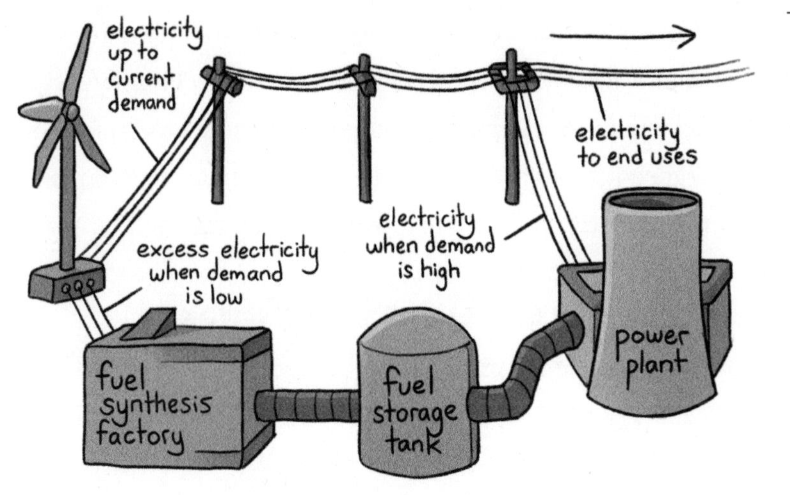

Fuel synthesis being used as a form of electricity storage.

Relatedly, any process—such as atmospheric sequestration—which uses a lot of electricity but can run at any time (or vary its power demand throughout the day) can balance out the grid demand with the intermittent solar and wind generation and constant nuclear generation.

These "flexible loads" will become more common on the grid as fuel synthesis and sequestration technologies scale up. Flexible loads could also include water desalination plants and other equipment not related to the energy system. The key factor in how flexibly these loads can operate is how cheap the capital cost is for the equipment in question. If it costs a lot to build a fuel synthesis plant, a company may only make its money back within the lifetime of the plant by operating it close to full time. If a plant can be built extremely cheaply, it might be economical to run it only a fifth of the time when there is a lot of excess electricity. Therefore, public funding for research and startups working on low-capital-cost flexible loads should also be a major part of a strategy to help solar, wind, and nuclear scale up as much and as fast as possible.

BASIS OF A CLEAN ENERGY TRANSITION

Because three of the other four pillars rely heavily on a plentiful supply of cheap, clean electricity, Pillar 1 needs to be carried out most immediately. Doing so will also bring the benefit of early large-chunk emissions reductions that limit later sequestration needs.

Government convening could jump start scale-up projects in nuclear using standardized designs based on roughly current technology. Policies such as state/provincial or national clean electricity standards, carbon pricing, subsidies, and simplified permitting regulations for the siting and construction of solar, wind, and nuclear plants can speed up the deployment of ready-to-go technologies.

With mass-manufacturing and policy efforts combined, and possible mandates for CCS, governments can guarantee the first wave of additional clean electricity generation is deployed. Efforts should aim to deploy enough clean equipment to at least replace current electricity generation levels by around 2030, on the way to generating the 100–150 PWh of electricity per year needed by 2050. Government-supported research can ensure that technology for all generation options continues making significant strides so that the remaining deployment can happen cheaply in the following twenty years.

6

PILLAR 2: ELECTRIFICATION

Most of the energy system could theoretically be decarbonized through electrification: converting equipment that currently runs on fossil fuels to equipment powered by electricity. We see this starting in our everyday lives with electric cars and electric home heating. The conversion will take place on a broader scale, too, in areas such as freight transportation and industrial processes.

VARIED EQUIPMENT

The technology required for electrification is different for every individual end use. To electrify personal transportation, an electric car motor and battery are both needed at capital costs that are lower than fossil fuel engines. An electric ship motor will call for a different design, and might run off the same or similar batteries, or off of a hydrogen or ammonia fuel cell (especially for longer-distance shipping). Electrification of trains requires building overhead power lines, a totally different challenge (those lines could also be used to extend and interconnect the grid). And heating for buildings requires improvement in the cost and cold-weather performance of heat pumps, which are like reverse refrigerators and bear no resemblance to engines or motors.

Government bodies should identify all the specific electric equipment that is either essential or would be highly useful, and convene researchers or industry players (depending how far along the given tech is) to ensure that at least one cost-competitive electrified option is available for each.

The good news here in Pillar 2 is that several electrified options are already close to becoming cheaper than their fossil competitors. Electric heat pumps are generally cheaper than oil heating and often competitive with other options depending on the climate, building's insulation, and other factors.[34]

Electric heat pumps don't work in extremely cold climates

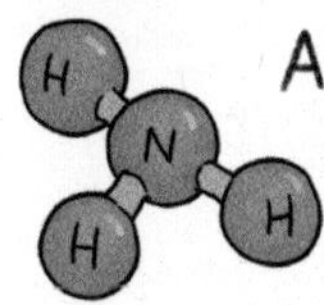

The main categories of equipment to focus electrification efforts on.

yet, but have improved to the point that they function in regions accounting for the vast majority of human population, and they can always be backed up by less efficient but equally non-polluting electric resistance heating (running electricity through lots of wire to heat up the wire and therefore the air around it). If clean electricity generation options outcompete fossil fuel electricity generation, the cost of electricity will by definition drop at least slightly, making heat pumps and other electrified equipment more competitive.

An air-source heat pump, with the compressor in the outside unit with the cold coil.

Electric cars have started to compete commercially, but their capital costs are still double those of equivalent gasoline cars, owing to the cost of the batteries.[35] If lithium-ion batteries came down further in cost, or if new, cheaper batteries were developed to replace them, electric car manufacturing could be scaled up to the point that electric cars quickly outcompete gasoline cars.

Trends toward self-driving cars and fleet ownership of cars (a Lyft/Uber-type company of the near future might own thousands of cars and bring you where you need to go on demand, reducing the number of personally owned cars) may also help promote the adoption of electric cars, which have far lower maintenance costs compared to gas cars (this is key when cars are operating like taxis, driving a much higher percent of the time, rather than sitting in individual people's driveways most of the time, as cars do now).[36]

The efforts required in this pillar span from basic engineering to mass-manufacturing scale-up. Government projects need to convene industry players to get certain key electric options designed. Policies and initiatives to scale up electrified equipment will help bring capital costs down to below or around those of fossil options. And improvement in technologies themselves, such as battery or heat pump chemistry, could help make electrified options definitively cheaper. Support for start-ups and existing companies that are conducting demonstration projects and bringing new designs to commercialization is key.

TIMELINE FACTORS

This pillar needs to be carried out relatively quickly. Electrified options may (easily, in some cases) become cheaper to install than installing a *new* fossil option for the same process. But it is much harder for these options to become so cheap that it makes economic sense to retire a still-functional piece of fossil equipment early and replace it with a new piece of electric equipment. Some options may get there, but for most, the deployment opportunity will be when each equivalent piece of fossil equipment is naturally retired. That turnover happens about every fifteen years for cars, but fifteen to twenty-five years (or longer) for home furnaces, and longer still for much industrial equipment. Therefore, electric options have to

Electrified options may (easily, in some cases) become cheaper to install than installing a *new* fossil option for the same process. But it is much harder for these options to become so cheap that it makes economic sense to retire a still-functional piece of fossil equipment early and replace it with a new piece of electric equipment.

become definitively cheaper in the next ten years or so in order to be ready at the moments when most relevant fossil equipment is being retired. Even then, there will be some portion of fossil equipment that doesn't get replaced naturally by 2050—so drop-in fuels (see Chapter 7), policy mandates or subsidies, or sequestration will be needed to fill in those gaps. Several recent reports estimated that 10–30% of equipment that could be electrified won't happen to get converted by 2050.[37]

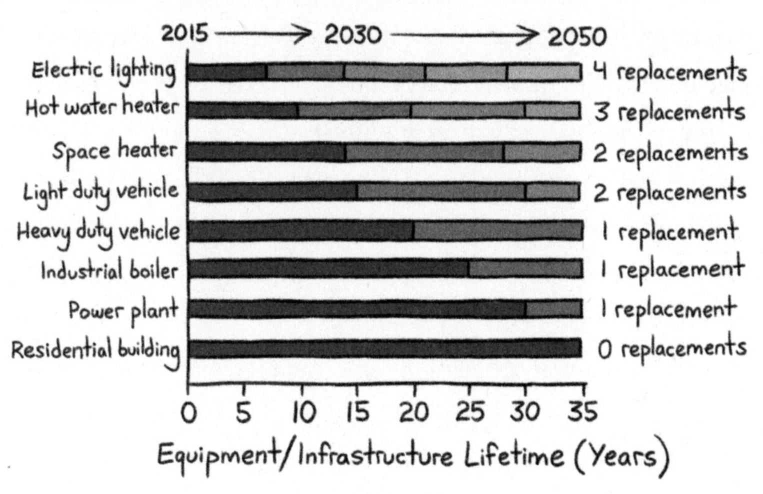

Most relevant equipment will only be naturally retired once or twice in the timeframe we have to achieve negative emissions.[38]

SCALING UP ELECTRIFICATION

This pillar is also one of the easiest to support with policy mandates or incentives. With several major electrified options nearly cost competitive, regions—even states and provinces in countries that aren't yet ready to act nationally—can subsidize the adoption of electric cars or air-source heat pumps through tax credits or rebates. Government programs can educate individuals and convene companies to encourage adoption of these technologies. Policies can also incentivize large companies to spur deployment of electrified options. For example, states can enact performance-based regulation of electric utilities in which utility companies make more money if a target number of charging stations, or electric cars, or heat pumps, are deployed in their service area within a given timeframe. Under such a system, utilities might choose to build charging stations themselves, or they might engage in more widespread education and encourage companies and families to adopt electric heating or other clean technologies.

Government purchasing power can also be used to directly scale up emerging electrified equipment by providing the first guaranteed market for various technologies. Most public building projects or public purchases of vehicle fleets are put out for contractors or suppliers to bid on. The bidder with the cheapest option that meets quality criteria gets hired—and criteria could include that the product has to be carbon neutral. Government purchasing departments, for example, could consider bids to construct new government buildings only if the design included air-source heat pumps and advanced insulation. Or states could put out a request for bids to supply electric vehicles for the entire state fleet as it gets replaced over the following years. The US federal government recently helped a new industry get off the ground in this way, by purchasing satellite launches from SpaceX and thus creating the first market for private space equipment.[39]

7

PILLAR 3: SYNTHESIZED FUELS

NEED FOR FUELS

So much discourse focuses on the mantra of "electrify everything," but in fact, some equipment will still rely on liquid or gas fuels in 2050. For example, we are unlikely to see an affordable long-distance electric airplane by that time due to the need for high-density energy carried on board.[40] And as noted at the end of Chapter 6, some fraction of fossil fuel equipment that could be electrified won't be replaced by 2050. Models tend to put maximum electrification potential at 40–60% of total end-use energy demand by 2050.

To fill the remaining need for fuels, fossil fuels can be used with CCS—but only on stationary equipment, not airplanes or other vehicles. And CCS will happen only where policy mandates it, as it is always an added cost. Perhaps every country will end up finding political will to mandate CCS, but for the sake of a 100% solution framework that can succeed *without* every single country taking dramatic action, we must assume CCS plays a limited role.

The second option is biomass—wood burned directly, crop waste (e.g., corn stalks), or dedicated crops (mostly people talk about perennial grasses)—the latter two probably converted into biofuels. As noted before, biomass plays a minor role in this framework because biomass-derived fuel can only be considered carbon neutral if the growing of the biomass itself doesn't produce more emissions than the biofuels save by displacing fossil fuels.

In theory, plants capture atmospheric CO_2 when they grow, and the same amount of CO_2 is released when they are burned (provided their production is powered by clean electricity or carbon-neutral fuels). However, growing such "energy crops" will require extra land and increase pressure toward deforestation (one of the largest sources of emissions, see Chapter 8).

Even with strong anti-deforestation and reforestation policy, strategic use of dense energy crops might only be able to serve something like twenty-five PWh of end-use demand per year, or less than half of the baseline projection of fuel required in 2050.[41]

Right now, even Europe—which generally has extensive regulation—counts much deforestation-causing wood biomass as "carbon-neutral." If even a few major countries don't adopt stringent regulations about sustainable sourcing of biomass ahead of 2050, other countries would have to try using tariffs or other policies to minimize the global market for unsustainable biofuels. If faced with that choice, which is fairly likely given the political difficulties of controlling biomass production, proactive countries might instead favor the approach of subsidizing carbon-neutral fuels that are synthesized from scratch to displace the market for both biofuels and fossil fuels. Activists worried about a scenario in which the world becomes reliant on biomass-derived fuels, and then sustainable growing standards become loosened and the whole system becomes carbon-intensive again, should also take a keen interest in synthesized fuels as a better guarantee of carbon neutrality. Directly synthesized fuels are guaranteed to be carbon neutral so long as we successfully decarbonize our electricity generation system.

CLEAN FUEL SYNTHESIS

Such synthesized fuels would be made by using clean electricity (or occasionally direct sunlight) to drive chemical processes. Most would start by splitting water into hydrogen and oxygen. The hydrogen could be used as a fuel, or could be combined with carbon split from captured atmospheric CO_2 or with nitrogen filtered from the air to form different fuels. The basic chemical reactions involved are the exact opposite of the ones that happen when a fuel is burned.

Burning jet fuel:

$$C_xH_y + O_2 \rightarrow CO_2 + H_2O + \text{energy}$$

Synthesizing jet fuel:

$$CO_2 + H_2O + \text{electricity} \rightarrow C_xH_y + O_2$$

CARBON FUELS

There are essentially two categories of synthesized fuels: carbon fuels and non-carbon fuels. Carbon fuels include gasoline, diesel, jet fuel, methane, perhaps methanol, and kerosene/propane and other minor fuels. Synthesized versions of these could consist of the exact same chemical(s) or some equivalent chemical that is easier to synthesize but functions the same way in the relevant equipment (e.g., a truck engine). Synthesized carbon fuels do emit CO_2 when burned, but it is the same amount of CO_2 that was captured from the air to synthesize the fuel in the first place. As long as the process is powered by clean electricity (or by other cleanly synthesized carbon or non-carbon fuels that were produced with clean electricity elsewhere in the world), the fuels are carbon neutral.

Generally, carbon fuels will be made either chemically identical to current fuels or similar enough that no new equipment is needed to run off them. They are called "drop-in" fuels because they can drop into existing infrastructure. This gives them a huge advantage in terms of rollout rate: once they can be manufactured more cheaply than whatever fossil fuels are still in use at that moment (perhaps with the aid of a manufacturing subsidy from one or more forward-looking countries), they could replace 100% of their fossil competitors nearly overnight—or as fast as the fuel synthesis equipment could be manufactured and connected to electricity sources. Carbon fuels are more "energy-dense" than non-carbon fuels or batteries, meaning a given volume of fuel can power a process for longer, so they are ideal for airplanes and freight transportation.[42] The fact that these fuels can be "dropped in" will allow them to eliminate emissions from the portion of individually owned furnaces and such that don't get converted by 2050 but are too numerous and small to make CCS a viable option.

The same type of chemical reaction design that is needed to develop affordable drop-in fuels could also be used to synthesize a variety of chemical feedstocks and other commodities (such as plastics or formic acid) that are currently derived from crude oil. This would help chip away at the market for oil drilling.

NON-CARBON FUELS

Non-carbon fuels include hydrogen and ammonia. Non-carbon fuels usually require conversion of end-use equipment in a similar way to electrification, and therefore also require that new end-use equipment be brought down in capital cost. In some cases, hydrogen or ammonia can be dropped into existing infrastructure. They can definitely be mixed in with fossil fuels up to a certain percentage in existing infrastructure, which doesn't get to 100% elimination of fossil fuels but could provide an early market while ammonia- and hydrogen-fueled equipment comes down in cost.[43]

The advantage non-carbon fuels have is that they are cheaper to synthesize than carbon fuels. Synthesis of non-carbon fuels skips the step of capturing atmospheric CO_2 and using electricity to split the carbon from the oxygen and reattach the carbon to hydrogen that has been synthesized from water. The hydrogen itself can be the fuel, or it can be bound to nitrogen (ammonia is NH_3), which is much easier to filter from air than CO_2 (nitrogen is 80% of the air while CO_2 is 0.04%). In either case, the energy requirement to synthesize the fuel is less than for carbon fuels, and therefore non-carbon fuels are likely to be cheaper per unit of energy output when they are burned. If the equipment to use them can be made low enough in capital costs, non-carbon fuels could be the cheapest option for many processes.

Non-carbon fuels will be most useful for applications that could be electrified but that would benefit from more energy-dense fuel—for example, freight trucks. Electric freight trucks might become competitive, but carrying ammonia or hydrogen in a tank and turning it into electricity in an on-board fuel cell to power the motor would allow trucks to go much farther between refueling (and to refuel much faster) than they could if the "fuel" were a battery and they had to stop for hours and charge the battery every couple hundred miles. Similar logic applies to ships—batteries might be fine for short-range ships, hydrogen might be used for medium-range ships because of its greater energy density than batteries, and ammonia might be used for long-range ships because of its greater energy density than hydrogen.[44]

Even electric cars could be powered with hydrogen tanks and fuel cells instead of batteries. The rest of the drive train would be the same, but if hydrogen fuel cell equipment could come down below the capital cost of batteries (and if hydrogen distribution and fueling station infrastructure were built out), the shorter fueling vs recharging time, and the considerably longer range, would give hydrogen cars a serious advantage.

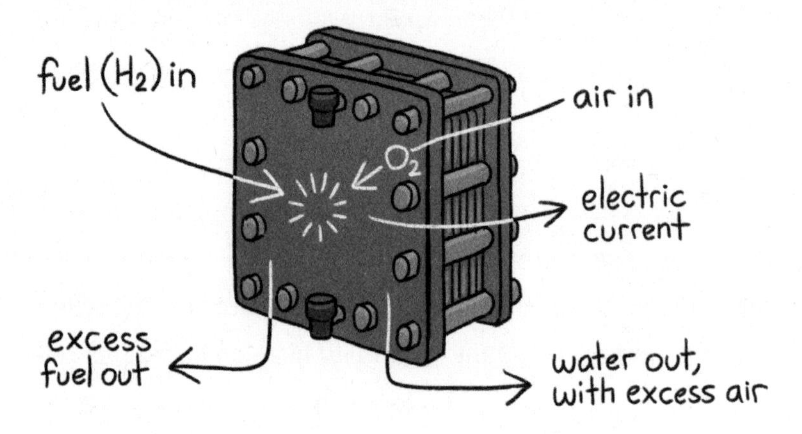

A hydrogen fuel cell.

HYDROGEN IS TOP PRIORITY

Because all synthesized fuels include hydrogen (or are pure hydrogen), a top priority for innovation efforts should be to lower the cost of clean hydrogen below that of methane-derived hydrogen. Right now, nearly all hydrogen used in industry is made by using steam to split methane into hydrogen and CO_2 and releasing the CO_2 into the air (methane also leaks into the atmosphere during extraction and distribution of the methane feedstock, see Chapter 8). Hydrogen can also be made through electrolysis, which uses clean electricity to split water into hydrogen and oxygen, and emits no greenhouse gases.

Making electrolysis-based hydrogen cheaper than methane-derived hydrogen requires lowering the capital costs of electrolyzers and the cost of input electricity.[45] New equipment, plant designs, or manufacturing practices could lower the cost of electrolysis equipment. Scale-up of clean electricity generation could provide cheaper clean electricity, especially during times of low grid demand.

Other proposals for clean hydrogen production include "gasifying" biomass into hydrogen.[46] Like all biomass efforts, caution to prevent additional deforestation is key (see above and Chapter 9), but biomass-derived hydrogen is less researched and optimized already compared to electrolysis and so might have more opportunities for cost reduction.

Producers could also split hydrogen from methane and use CCS to store the CO_2 rather than releasing it into the atmosphere. Like all CCS, this is an extra cost compared to the same process without CCS, so it will only happen where policy mandates it, making it a strategy that can't be counted on globally.

Getting the capital cost of electrolyzers down is therefore the top priority for the synthesized fuels pillar. Cost-competitive clean hydrogen is a precursor to any other synthesized fuel being competitive. Once clean hydrogen can be produced more cheaply than fossil-derived hydrogen, various existing processes can convert it relatively cheaply into any number of fuels, carbon or non-carbon. The hydrogen itself tends to be the largest cost in any of these clean fuels.[47]

ROLE OF CARBON FUELS

There are a range of perspectives on the utility of synthesized carbon fuels. Some startups and academic labs see them in the context of incrementalist policy—a business or research opportunity to serve markets where a state or federal government has mandated lower-emission vehicle fuels. These niche-market approaches might provide the first steps to building out synthesized fuels toward cost-competitive manufacturing scale, but won't necessarily lead to much of a solution.

A couple of advocacy organizations with high hopes for nuclear power's cost reduction through iteration suggest that carbon drop-in fuels should be thought of as the entire solution for energy use emissions.[48] They promote the idea of building out nuclear electricity generation for almost the sole reason of powering fuel synthesis processes, with the hope of achieving such low costs of drop-in fuels that they outcompete fossil fuels everywhere. The target cost would have to be something like $1.00–1.50/gallon for gasoline and diesel, because oil companies can reduce prices significantly while still making profits.[49] Those targets for synthesized drop-in fuels are probably not achievable within thirty years, but if electricity could truly become far cheaper and electrolysis capital costs could come down dramatically, this approach would be possible. The advantage would be a much faster transition—dropping in clean fuels to *all* existing fuel-consuming equipment—than electrification-heavy, mandate-focused, or non-carbon fuel routes.

That small possibility alone makes it worth pursuing and funding vigorous research and engineering efforts for drop-in fuels. We will still benefit from more modest cost improvements even if we don't bring drop-in fuels to the point that they outcompete fossil fuels overnight.

Most likely, with serious innovation carbon drop-in fuels can come down in cost to a point slightly more expensive than their fossil equivalents. Then, governments will have to subsidize their use (or price fossil fuels appropriately) to fill in the gaps that electrification can't cover. Synthesized fuels—and, to some degree, sustainable energy-crop biofuels—have the advantage that they could be subsidized on the production side and sold into the global fuel market at costs cheaper than fossil fuels. This means one country could influence every country to use the cheaper carbon-neutral fuels—a key difference from CCS (and from broader biomass use) where every country must adopt individual regulation.

ROLE OF NON-CARBON FUELS

As noted, non-carbon fuels should be able to achieve lower costs per unit of energy than carbon fuels, so once clean hydrogen can be produced more cheaply than methane-derived hydrogen, non-carbon fuels could play a larger role than carbon drop-in fuels or fossil fuels even without policy intervention. They face a challenge akin to electrification, which is lowering the capital cost of equipment to run on them.

Ammonia is synthesized from hydrogen—currently fossil-derived hydrogen with an energy-intensive synthesis method, but both the hydrogen source and the synthesis method can be changed. Technologies exist to split ammonia back into hydrogen and nitrogen so that the pure hydrogen can be used. The advantage of ammonia is that it becomes a liquid much more readily than pure hydrogen, so it is easier to transport from the point of synthesis to the point of use (or to carry as a fuel on a long-distance ship or truck).[50] The advantage of hydrogen is that we already have equipment that can power various processes using it—hydrogen fuel cells are relatively established, though expensive. Ammonia fuel cells (or combustion engines for either ammonia or hydrogen) need more development to be ready to roll out globally.[51] Ammonia, like fossil fuels, is toxic, and happens to smell particularly bad, so it makes sense to use it directly in industrial processes and freight transportation but to turn it

electricity
nitrogen (from air)
hydrogen (H_2)
ammonia (NH_3)
water
Ammonia can be used to transport hydrogen over long distances, where it can then be used for industrial heating or for fueling cars.
NH_3
Ammonia
Nitrogen (released)
Hydrogen
NH_3
H_2 FUELING STATION
for heating
for fuel

back into pure hydrogen for use in personal equipment such as cars and home heating systems.

Note that once consumed in a fuel cell, both hydrogen and ammonia (or various carbon fuels which can also be used with fuel cells—methanol is a prime prospect) are converted to electricity. The end-use equipment therefore consists of the fuel cell itself and then the same or similar electric equipment as would otherwise be powered by the grid (if stationary) or by a battery (if in a vehicle). Ammonia and hydrogen can also be burned in engines or turbines to drive equipment without electricity as an intermediary, and those pieces of equipment require designing, testing, and demonstrating to become viable.

FLEXIBLE LOADS

Synthesized fuels offer an additional benefit beyond the fact that they can fill in gaps left by electrification and could drive a very quick decarbonization of energy systems if they become cheaper than expected. As discussed in Chapter 5, one problem with clean electricity costs is that the output of solar and wind fluctuates uncontrollably. At the same time, nuclear plants almost always run near peak output and are currently hard to ramp up and down. Neither option exactly matches grid demand, which fluctuates over the course of a day and over the course of a year semi-predictably. Fuel synthesis plants could provide a "flexible load" on the grid—they could ramp their electricity consumption (and therefore fuel output) up and down, or turn on and off entirely, to balance the difference between clean electricity generation and the instantaneous demand of the grid. Equipment with high capital costs can't serve this role, because it is only economical to run such equipment most of the time to produce a product to sell and make back the capital investment. For that reason, it is especially worthwhile to invest research and development money and effort into potential synthesis technologies that could be super-low capital cost, making them affordable to operate only a small percent of the time.

By turning excess electricity into fuel for transportation and industry, the synthesis of fuels can also help integrate currently distinct sectors. Fuels can become a sort of translator between the needs and demands of different sectors. They could do the same between different regions, as liquid and gas fuels are among the easiest forms of energy to transport across large distances, or between times of year, as such fuels are easy to store for months.

This "fungibility" across sectors, regions, and times can lower overall costs by limiting the need for overbuilding generation or capacity in any one sector or location. Policies and technological systems to better enable coordination across sectors (for example, timing car charging based on intermittent electricity generation) can also help lower costs in this manner.

SCALING UP SYNTHESIZED FUELS

Overall, synthesized fuels could end up being responsible for anywhere from 5% to 40% of global emissions reductions, depending on how cheap they become at what times.[52] Although left out of much political discourse, synthesized fuels will be required for some part of a 100% solution. And they are a prime target for concerted, coordinated research funded and convened by government bodies. This pillar requires the most basic research funding and coordination, because chemical processes need to be developed and tested. Commercializing synthesized fuels will also require support for demonstration, startup efforts, and initial scale-up in order to improve manufacturing techniques, business practices, and capital costs of end-use equipment.

Policies that use government purchasing power to guarantee a market for the first batch of synthesized fuels, subsidize or mandate their use more broadly, or incentivize corporate players to spur their adoption—for example, providing a tax credit for companies that switch fully to carbon-neutral fuels (or electrified options) for their heating and transportation—could all speed the scale-up of synthesized fuel technology.

Synthesized fuels also rely on cheap, clean electricity—Pillar 1—to be viable. The success of synthesized fuels will depend on the combination of how cheap various components, especially electricity and hydrogen electrolyzer capital costs, become.

8

PILLAR 4: NON-ENERGY SHIFTS

Over a third of emissions come from processes other than burning fossil fuels, and yet they don't get nearly as much attention as fossil fuel emissions. Non-energy emissions come from three sectors: agriculture, industry, and wastes.

AGRICULTURE

About 20% of the world's emissions are related to growing crops or raising livestock to feed everyone in the world. Some of these emissions come directly from the soil and livestock, and a large portion come from the conversion of land to make space for them.

DEFORESTATION

The biggest slice of agricultural emissions, making it the source of around 10% of total global emissions, is deforestation. To make more land area available to graze livestock or grow crops, farmers regularly cut down tropical forests which have been storing CO_2 in their plant matter and soil for a very long time. The one-time emissions from cutting forests (which are either released right then if the forests are burned, or over a short amount of time if the materials are left to decompose) are large. Then, without those trees to sequester more CO_2 from the atmosphere, a less immediate form of

> **Over a third of emissions come from processes other than burning fossil fuels.**

emissions occurs by reducing the annual amount of carbon sequestered by forests. This reduction is significant, but hard to quantify among annual emissions as it continues indefinitely without an objective "starting point" to compare any given year's total amount of global tropical forest to.[53]

Tropical deforestation contributes the bulk of this slice of emissions. Tropical forests store much more CO_2 in their dense growth and are often much older than temperate or boreal forests. However, forests are not the only relevant ecosystem—when this book refers to "deforestation" it also includes what the Intergovernmental Panel on Climate Change calls "other land use change," encompassing the draining of peat bogs and loss of certain coastal ecosystems such as mangrove swamps. These other carbon-rich ecosystems can also fall prey to the need for more agricultural land and cause a portion of what we're calling "deforestation" emissions.[54] They are also just as good targets for "reforestation" (or "ecosystem restoration," to be more accurate) as a sequestration method—in fact, recent reports have suggested that they have a stronger short-term potential for sequestering carbon than forests, and should be the prime initial target for ecosystem restoration.[55]

Not all ecosystem loss is caused by agriculture, but the bulk of climate change–relevant land change is. The culprits are a mix of small-scale subsistence farmers and large-scale commercial farmers, in roughly similar proportions depending what region of the world one looks at.[56] There is virtually never a malicious intention: people simply need money to survive, and growing food is one of the only ways they can make money, so they make the space to do so. Clearing ecosystems is cheap, and even if the soil becomes unhealthy and crop yields decline in a few years, growing a cash crop for that time can be well worth it to many farmers around the world. The greatest drivers of tropical deforestation are raising cattle, growing soy (some of which becomes feed for the cattle), and growing palms for palm oil.[57]

Government regulations are particularly effective in limiting deforestation. Brazil did this successfully for about a decade until deforestation rates started increasing again around 2014 (and have since spiked significantly after anti-environmentalist president Jair Bolsonaro took office in 2019).[58] Other solutions involve practices that reduce demand for land and therefore reduce pressure toward deforestation. These could include better crop rotation, cover crops and interspersing of crops and livestock, denser livestock grazing practices, use of greenhouses (which extend the growing season and improve crop resistance to heat waves or pests), "vertical farming" (indoor growing in tall columns with plants densely packed, artificially lit using clean electricity, and maintained by automated machines in the building), and other techniques that achieve the goal of growing more from a given amount of land.[59]

Vertical farming might look something like this, with columns of growing greens packed close together.

Government intervention to shift economies toward denser farming or non-agricultural sectors and enforcement of policies against deforestation can play a significant role in reducing this slice of emissions. Leadership from key tropical countries, and from other national leaders who can form partnerships with them, may be key to seeing progress in reducing deforestation. A relatively small number of countries contribute most of the deforestation emissions in the world, with a hugely disproportionate 45% of the total coming from Brazil. Indonesia is next at about 9% and about twenty more countries contribute around 1–3% each.[60]

Shifting diets away from meat, especially meat from livestock, could also reduce deforestation. Grazing cattle takes up far more land area per unit of nutrition that someone will eventually eat compared to growing food crops. On top of that, cattle are usually fed with crops (corn and soy) that could have fed humans directly. The growth of those feed crops requires yet more land area.[61]

Policies that discourage the consumption of livestock—for example, carbon prices that incorporated surcharges on emissions-intensive goods, including beef—might slightly reduce global demand for livestock and thereby reduce the market for converting forests to grazing land.

CATTLE METHANE EMISSIONS

Livestock farming's role in driving deforestation represents its largest contribution to emissions, but it creates emissions in another way as well. Cows, goats, and sheep belch methane as they digest food, and methane is a greenhouse gas eighty times stronger than CO_2 as measured over the ten- to thirty- year timeframe we have to keep temperatures from increasing more than 1.5 degrees Celsius (it's actually closer to a hundred times stronger, but only stays in the atmosphere for about ten years, whereas CO_2 stays for thousands of years, so methane is often measured as thirty times more powerful over one hundred years, or eighty-four times more powerful over twenty years).[62] Enteric fermentation (the process of methane being created as these ruminants digest food) is the second-largest source of agricultural emissions behind deforestation, equal to about 5% of the global greenhouse gas total.[63]

Reducing the demand for livestock meat (and milk, which is much more carbon intensive than plants but much less than beef) could reduce these methane emissions. But in this case, a technological solution also exists. Several studies in recent years have shown that a certain species of seaweed, when mixed into cattle feed, dramatically reduces the cows' methane output—some studies show reductions of 99%. There are several startups working to scale up production of this seaweed and commercialize its use in the near future.[64] Investment in such efforts and in projects to educate farmers about using these products in their cattle feed could virtually eliminate the livestock methane slice of greenhouse gas emissions.

FERTILIZER EMISSIONS

That leaves us with crops. To grow plants, farmers have to add some kind of fertilizer to the soil. Fertilizer, whether

organic (compost or manure or similar) or synthetic (urea or other ammonia-based compounds) supplies nitrogen to the growing plants. The addition of nitrogen-dense fertilizer has enabled the vast growth in agricultural production over the past hundred years, and is one of the greatest forces countering pressure toward deforestation. Applying fertilizer means more can be grown per hectare of land, meaning less land area is needed. However, all fertilizers also cause emissions, because some of the nitrogen spread on the soil becomes nitrous oxide, another greenhouse gas that is hundreds of times stronger than CO_2, which spreads into the atmosphere.[65]

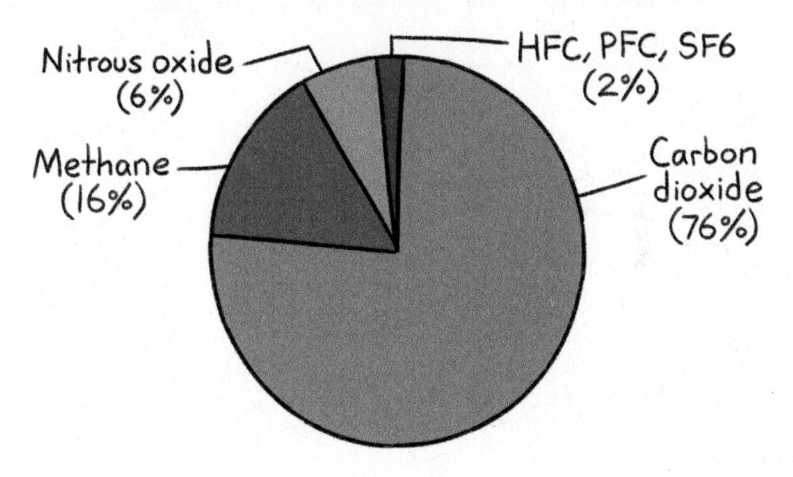

Global emissions by type of greenhouse gas (based on the standard comparison over one hundred years—methane would look more important if compared over ten or thirty years).[66]

Some farmers wrongly assume that more fertilizer means more yields. This is only true up to a point, after which extra fertilizer may actually decrease yields, while increasing nitrous oxide emissions.[67] So government or nonprofit programs could educate farmers about how to use fertilizer more efficiently to decrease, but not eliminate, these emissions. This is one slice that will need sequestration to make up for it by 2050, as these emissions cannot practically be eliminated while still growing enough food for the world.

SOIL EMISSIONS

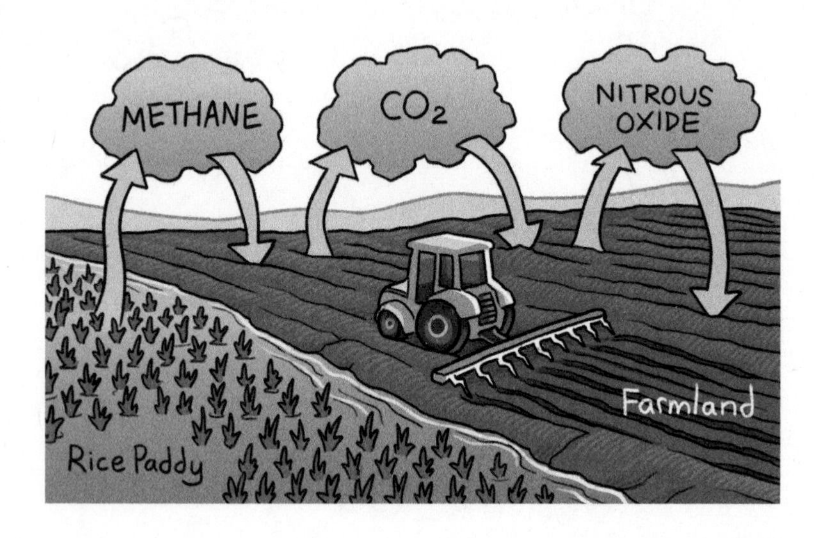

Other crop-related emissions include those from the soil itself. Various chemical processes, largely inside bacteria that live in soils, both produce and consume greenhouse gases. For instance, in wet soils (such as rice paddies), anaerobic bacteria tend to produce methane. Dry soils may be a sink (sequestration site) for methane. CO_2, methane, and nitrous oxide are all produced and stored in soil, with the balance depending on how wet the soil is, its temperature, what kinds of crops (or grasses, or forests) are being grown on it, and more.[68] Some of these emissions can be mitigated by managing soils differently, which may also contribute the first few gigatons of sequestration needed (see Chapter 9).

There has been long-developing research into whether tilling soil is good or bad for climate change. For decades it was thought that tilling stirred up and released CO_2 stored in the soil, and that no-till agriculture practices would sequester more CO_2. More recently, researchers have noted that most studies examined only the top layer of soil (about a foot deep), and while the results are true for that layer, tilling actually mixes CO_2 into the slightly deeper soil, thereby sequestering about as much as it loses from the top layer.[69]

OTHER AGRICULTURE EMISSIONS

Finally, small amounts of agricultural emissions come from the decomposition of crop residues and manure, as well as from energy use in agricultural processes.[70] The latter emissions will be eliminated through either electrification or synthesized fuels (see Chapters 6 and 7). The former may be reduced through increased use of composting or anaerobic digesters, which at least provide productive use of such crop or animal wastes, but some of these emissions will also likely remain.

In all, we can expect about half of agricultural emissions to be eliminated by 2050. Those from fertilizer, which are mostly unavoidable (and which prevent larger deforestation emissions if food were grown less densely without fertilizer), some portion of deforestation emissions (which would require strict regulation and enforcement in every country to eliminate),

and small portions of soil and other emissions, perhaps as much as 10% of global total emissions, are likely to remain in 2050. Agriculture emissions are the hardest to totally eliminate, but they are also the easiest to balance with farm-based and forest-based sequestration of greenhouse gases, as we will see in Chapter 9.

GLOBAL OUTREACH INITIATIVE

All agriculture emissions can be reduced through changes in agricultural practice—for example, growing more food per hectare of land to reduce pressure toward deforestation, or managing crop rotation, cover crops, and sharing of land between crops and livestock to build soil carbon and reduce fertilizer use. A global outreach initiative should be created to get as many farmers as possible to adopt these practices. Such an initiative—working on the ground with every farmer in the world—could be carried out by a private nonprofit organization, perhaps with public funding support. Or it could be run as a public program akin to the Peace Corps, with one or more countries funding, training, and sending volunteers around the world to work with farmers.

The most important and viable practices to adopt will vary from location to location, but in general crop rotation (growing two or more different kinds of plants in succession in the same

> **A global outreach initiative should be created to get as many farmers as possible to adopt these practices.**

plot of land to balance the nutrients taken from and added to that soil) is known to improve yields and decrease demand for fertilizer.[71] Grazing livestock in more efficient ways, such as interspersed with certain crops, could reduce pressures for deforestation.[72] Public funding and direction should support not only the outreach project to get practices adopted, but also research efforts to quantify which practices will have the most impact in which locations.

The outreach initiative will not only reduce emissions, but contribute the first few gigatons of carbon sequestration by promoting ecosystem restoration and soil carbon sequestration (see Chapter 9). In connection to the initiative, national and local governments in key countries (especially those with tropical forests) should enact policy favoring forest growing or accelerating the adoption of new agricultural practices, and should implement stricter and more consistent enforcement of anti-deforestation laws. Where viable, partner government bodies could also create policies that reduce food waste, which contributes both to unnecessary agriculture emissions and to direct landfill methane emissions (see "Waste" section later in this chapter).

INDUSTRY

Industrial processes use a lot of fossil fuels for energy (mostly to heat raw materials in manufacturing), but some also involve chemical reactions that release greenhouse gases unrelated to energy inputs. These appear in the pie chart as non-energy "industrial process emissions."

CEMENT

In the course of producing cement, limestone is heated in a kiln to turn it into lime, off-gassing CO_2 in the process. Lime, a chemical subset of limestone, makes up over half the material in the final cement. Most of the emissions from cement production don't come from the fossil fuels burned to power the process: half to two-thirds are given off during this step of converting limestone to lime.[73]

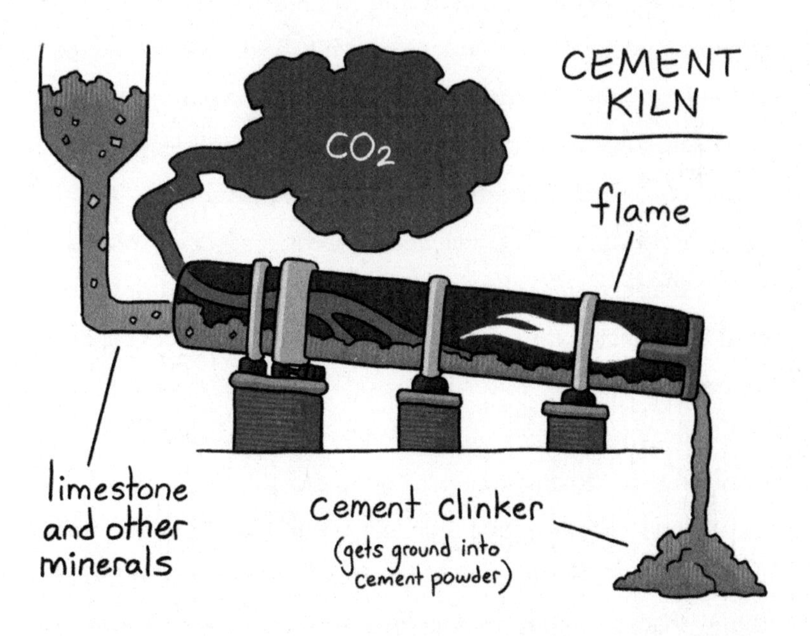

Over the years, various academic researchers have developed alternative cement production processes, which either off-gas less CO_2, or lock the CO_2 into the cement material itself during the process, thereby preventing most or all emissions into the atmosphere. A couple such processes are now being demonstrated commercially, such as one that reduces the proportion of lime in the final product, thereby reducing the direct emissions from turning limestone into lime, and which also requires lower temperatures in the cement kiln, thereby reducing the amount of coal used to heat the kiln. At least one company is commercializing cement that is hardened into concrete by absorbing CO_2 (rather than being hardened with steam as most cement is today), which further reduces the total emissions. That startup claims processes for hardening concrete could eventually lock up more CO_2 than is released in the cement production. While all concrete exposed to air absorbs small amounts of CO_2 over decades, concentrated streams of CO_2 (from factories with CCS or filtered from the atmosphere) could be injected into and sequestered by this kind of concrete in larger amounts.[74]

Other new processes can be designed to reduce the amount of fuel needed to create cement, or to reconfigure cement kilns to run on hydrogen for their heat input. Government efforts to convene academic researchers and bring more minds to the problem, to support startups, and especially to test out the most promising ideas at scale (proving new processes' reliability is key to convincing companies to switch over to them) will be crucial to ensuring that most cement production is decarbonized by 2050. Some of the near-term reductions in cement emissions may also need to come from mandated or incentivized use of CCS.

STEEL

Like cement, direct emissions come from the process of steel production, where coking coal is added to raw iron ore to "reduce" it to pure iron. The carbon from the coal combines with the oxygen from the iron ore and becomes CO_2, about two tons of the greenhouse gas being generated for every one ton of steel produced. This "reduction" process can be accomplished with hydrogen instead of coal (as can the heating of the steel furnace), and the necessary input of carbon to turn the iron into steel could come from charcoal (which would have to be sourced from sustainably harvested wood) instead of coal. One Swedish company is working to commercialize hydrogen-reduction iron. One Massachusetts company is working to commercialize a different process which uses electricity and catalysts at high temperatures, rather than hydrogen, to reduce iron ore into iron.[75]

About a third of the world's steel furnaces already run on electricity ("electric arc furnaces") but those are generally only used to process scrap steel rather than raw ore, and may not have large potential to replace traditional blast furnaces in the next thirty years.[76] Policies to encourage greater use of scrap steel could help drive the electrification of more steel furnaces.

Large-scale demonstration of alternate processes or pieces of equipment for both cement and steel production will be essential. Because these industries are massive, longstanding, and deal in structural building materials, they have little tolerance for risk in their processes. New processes must be proven thoroughly so companies know their material quality won't be compromised if they switch over. Wherever policy mandates or incentives can push it, CCS on both steel and cement plants can also eliminate the process and heating emissions, though some technology development might be necessary to make it affordable.

Scaling up the use of other building materials can displace some demand for cement and steel and reduce—but not eliminate—these emissions. Wood, for example, sequesters carbon in a building itself when it is sustainably grown and harvested.

Recent developments in cross-laminated timber make it possible to build even skyscrapers out of wood, often with equal or better strength characteristics compared to steel.[77] Growing a market for construction wood could help incentivize keeping forests as forests and managing them sustainably rather than clear-cutting them to grow soybeans for a few years until the soil is depleted.

AMMONIA/HYDROGEN

Beyond cement and steel, a couple smaller sources of industrial non-energy emissions stand out, including ammonia.

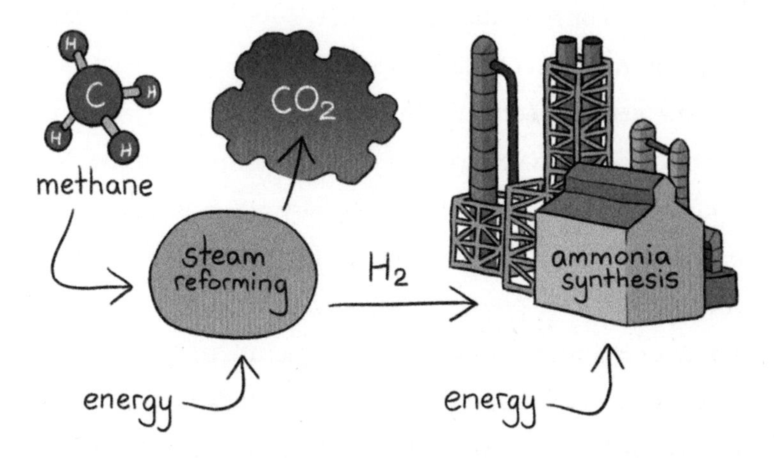

The synthesis of fertilizer (mostly ammonia) itself is energy intensive, but the direct emissions come when methane is split to get hydrogen as a feedstock for the process. Ammonia production happens to be the largest industrial use of hydrogen, but as noted in Chapter 7, almost all hydrogen currently used in industry has the same problem: methane (CH_4) is split using steam into hydrogen (H_2) and CO_2, and the CO_2 is released into the air. Hydrogen can also be produced by using electricity to split water, which does not release CO_2, but which is currently more expensive than "steam reforming" methane.

As discussed in Chapter 7, innovations to reduce the capital cost of electrolysis equipment may be the most significant improvement needed, as the up-front cost for electrolyzers can be larger than the total cost of methane-derived hydrogen for several decades. Other clean hydrogen produced from sustainable biomass or from methane with CCS could also decarbonize the ammonia synthesis process.

Some new ammonia production technologies show promise for replacing the entire current pathway and in some cases synthesizing ammonia directly from water and air without separately creating pure hydrogen.[78] However, clean hydrogen will still need to be produced not only for ammonia production and other industrial uses of hydrogen as a feedstock, but also for use as a carbon-free fuel itself and as a feedstock for the synthesis of carbon-neutral drop-in fuels (see Chapter 7). This should be a key priority for enabling synthesized fuels, and could benefit from government convening and coordination, extra research and scale-up funding, information sharing, and testing and demonstration initiatives.

ALUMINUM AND OTHER SOURCES

Finally, aluminum production, which is one of the few major chemical manufacturing processes currently powered by electricity, has some direct emissions from CO_2 being off-gassed by the electrodes as they operate.[79] New electrodes could be invented that use different materials and don't have the same side reaction that currently creates CO_2. Basic research funding

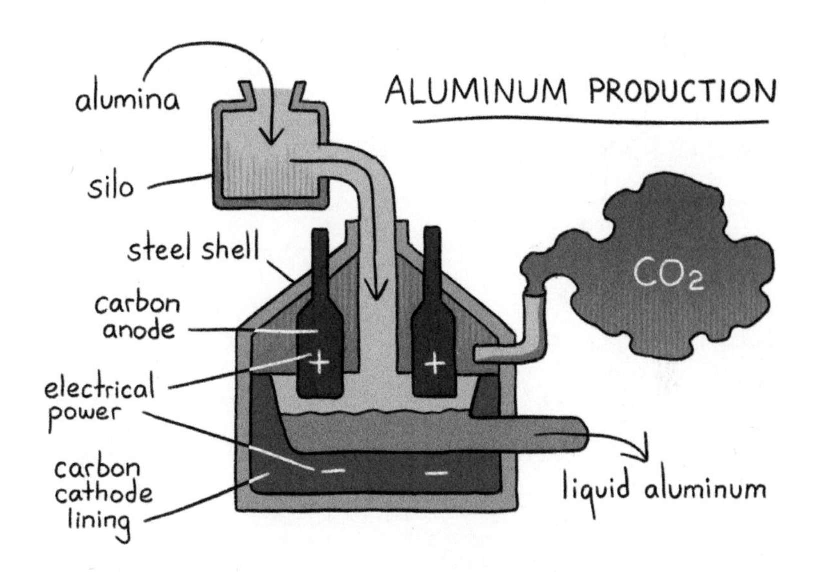

and then prototyping and testing-phase support are the key efforts needed for aluminum.

Still other direct emissions come from tinier industrial sources, but those explored here are the most important and readily fixable ones. The others, too, will need similar new processes developed, or in some cases may be small enough that we can rely on sequestration to make up for them.

WASTES

Agriculture accounts for almost 20% of total emissions, and industry direct emissions amount to another 7% or so. The final non-energy category includes material waste systems (contributing about 3% of emissions) and wasted "fugitive" methane (contributing about 5%).

BIOMASS AND WASTEWATER

Some emissions are released by wastewater treatment, and some from landfills. In particular, organic matter (food or farm

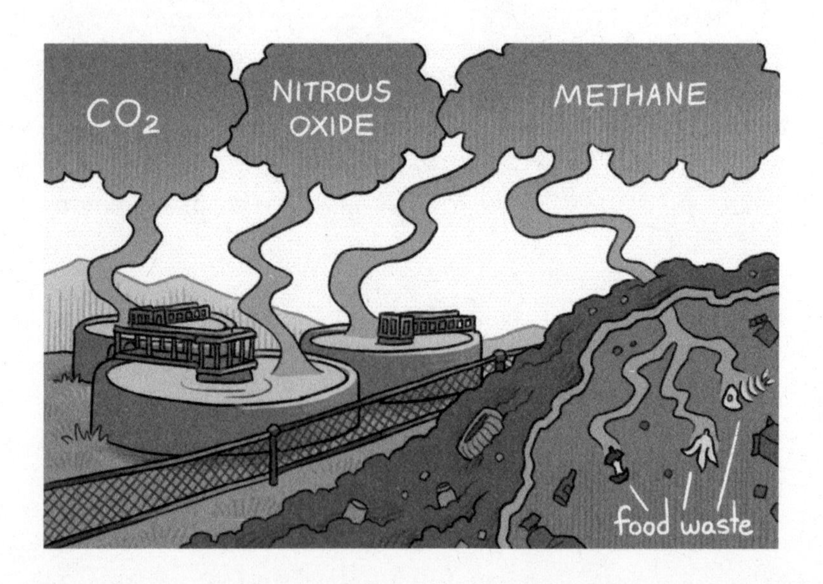

waste) that gets thrown out into landfills decomposes into methane, some of which leaks into the atmosphere.

Policy to reduce food waste could cut down this slice of emissions, and increased use of composting, anaerobic digesters, and other non-landfill methods for processing organic waste could reduce those methane emissions. Methane can also be captured from landfills and burned, which at least converts it into CO_2, a weaker greenhouse gas than methane itself. Through policy, the plants that burn this waste methane could be required to use CCS and make that process truly carbon neutral.[80]

FUGITIVE METHANE

A different category of waste emissions is "fugitive" methane that escapes from the energy system. When methane is extracted from the ground (and sometimes when coal and oil are extracted, even if methane is not also being intentionally extracted) some bit of it leaks into the air. When it is transported to end-use equipment through networks of pipelines, a little more leaks out. And in various bits of equipment along the way, methane is used for mechanical processes or partially burned in compressors to maintain pressurized pipelines, and the unburned methane is released into the air. Sometimes methane pipeline leaks cause explosions or fires in cities, but usually the methane simply floats into the atmosphere.

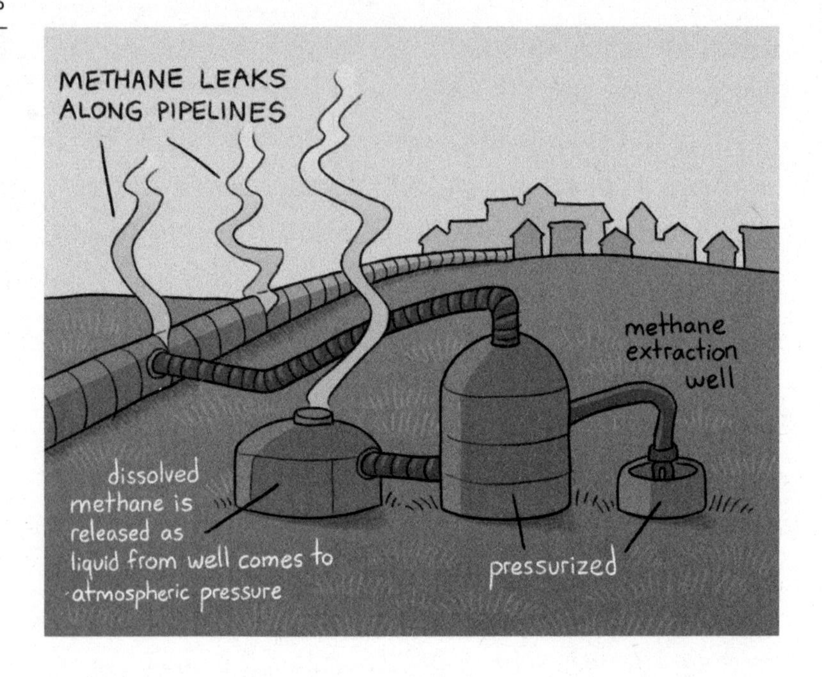

Methane being such a stronger greenhouse gas than CO_2, these fugitive emissions account for a full 5% of the global total of emissions.[81] They come from the energy industry, but the leaks are a total mistake, not a result of burning fossil fuels for energy, and the equipment-based emissions are basically accepted but not intended. They are one category of emissions that comes with absolutely no benefits to anyone (in fact, they mean economic losses for the methane extraction and distribution companies, meaning slightly higher costs for consumers). Tightening pipelines and other methane infrastructure, switching to electric compressor pumps, and adopting new equipment at wellheads could reduce these emissions. Switching to synthesized methane (see Chapter 7) could eliminate the half of fugitive emissions that come from extraction. And eliminating methane use altogether through electrification and conversion to other synthesized fuels would of course eliminate these emissions, though it's well worth pursuing policy mandates or business campaigns to reduce fugitive emissions in the meantime, because that can cut out a slice of emissions much sooner than we can expect all methane use to disappear.

REMAINING EMISSIONS

Including wastes, industry direct emissions, and agriculture, there are portions of emissions slices that probably won't be eliminated by 2050. Sequestration will have to make up for this remainder, which will probably be something around 15% of our current emissions (maybe 2% from material waste, 1% from pipeline methane leaks, 2% from industry, and about 10% from agriculture, with possibly tiny bits of energy-sector emissions also remaining).[82]

In order to keep eventual sequestration costs reasonable, this means that it is all the more important for energy system emissions to be virtually eliminated.

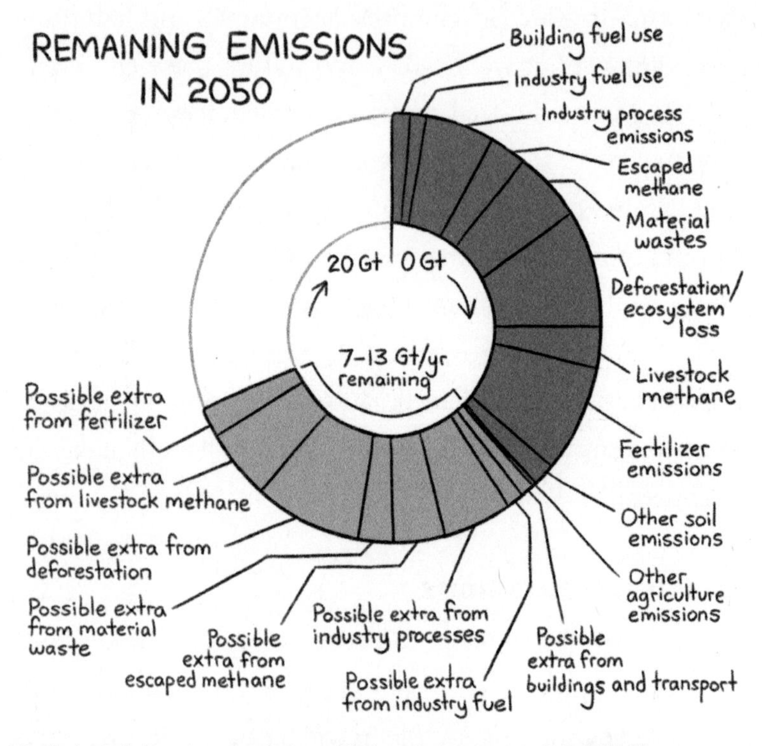

The dark shaded section contains most likely emissions to remain in 2050. The light shaded section contains possible additional remaining emissions, including extra amounts from many of the same processes. We can expect somewhere between 7 and 13 gigatons per year of remaining emissions around 2050. The full circle in this graph is 20 gigatons per year—the upper goal for sequestration as we will see in graphs throughout Chapter 9. [83]

9

PILLAR 5: SEQUESTRATION

To achieve negative emissions and stop the increase in global temperatures, the world will have to employ various carbon removal (or "carbon sequestration," or in this book simply "sequestration") methods to remove CO_2 from the atmosphere and make sure it doesn't get back there for a long time. There are six main proposed methods. Some are entirely nature-based, some entirely engineering-based, and some a blend of both.

The main proposals, detailed below, are growing forests, soil management, bioenergy with carbon capture and storage, direct air capture, enhanced weathering or ocean alkalinity enhancement, and ocean fertilization or ocean afforestation.

As we know from the earlier pillars, some sequestration will be needed to make up for remaining emissions in 2050. Likely about ten gigatons (Gt) per year of CO_2–equivalent emissions will remain even if we reach the goals of Pillars 1–4. And so ten gigatons of atmospheric CO_2 will have to be sequestered each year to make up the difference to net-zero emissions. Additional sequestration on top of that will take us into negative emissions. How fast negative emissions go will depend on how much countries are willing to pay each year to carry them out. The cost of sequestration is expected to range from nearly free for certain farm-based methods to around $100 per ton or slightly more for the purely technological methods, once implemented at scale.[84] The sequestration potential of nature-based methods, as we'll see below, will probably be nearly or entirely exhausted in making up for remaining emissions. So the negative emissions portion of sequestration will probably cost nearly or about $100/ton, at least for the next few decades. Assuming, based on the size of the global economy ($90 trillion/year, of which probably under 1% is politically viable for sequestration costs), that countries may in aggregate find the political will to pay hundreds of billions, but not trillions, of dollars, each year for total sequestration (both making up for emissions and achieving negative emissions), the negative emissions we can expect starting around 2050 might be 5–10 additional gigatons per year.[85] Depending on how cheap technologies get and how many tons of remaining emissions actually need to be made up for, the cost could be lower or the number of politically viable gigatons per year of sequestration could be higher.

As we'll return to briefly in Chapter 10, to bring temperatures back to historic averages sooner than the end of this century, eventually 20–50 Gt per year of total sequestration

How fast negative emissions go will depend on how much countries are willing to pay each year to carry them out.

might be necessary, which speaks to the importance of understanding and developing all the possible sequestration methods as soon as possible. For now, we'll compare sequestration methods based on how many gigatons they might be able to sequester per year in the near future, with the nearer-term 10–20 gigatons per year total goal. Estimates of "viable" annual sequestration amounts vary considerably. Presented here are the best estimates available for how much could be sequestered each year without unreasonable policy expectations and for less than or about $100 per ton of CO_2.

GROWING FORESTS

Forests—and, as discussed in Chapter 8, wetlands and other ecosystems—lock up large amounts of CO_2 both in the trees and other plants above ground and in the soil itself. Just as deforestation is one of the major causes of CO_2 emissions, reforestation or ecosystem restoration (growing forests where deforestation has occurred, or restoring other carbon-rich ecosystems) and afforestation (growing forests where there weren't any recently but could be) can drive significant amounts of carbon capture by plants, and therefore contribute significant negative emissions. Restoring ecosystems is cheap, especially compared to other sequestration methods, and comes with co-benefits—availability of some amount of lumber that can be sustainably harvested, habitats for species that we either rely on or want to protect for their own sake, beauty to enjoy from a distance or on hikes, and improved air quality in regions near forests.

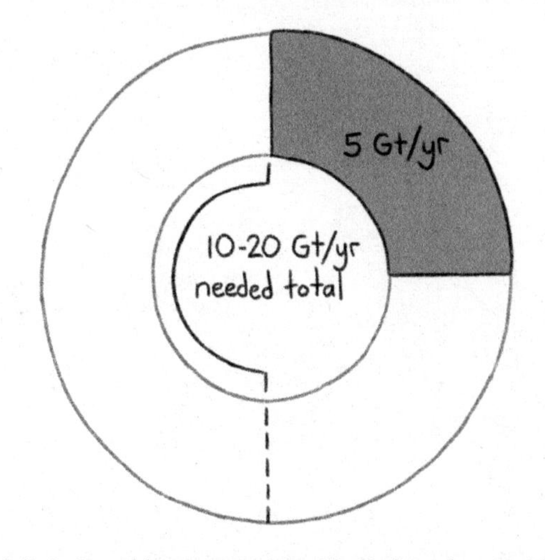

One of the main limitations of this sequestration method is that only so much land area can be turned into forest each year—particularly with continued pressures to cut down forest and use the land for crops or livestock. Estimates put the upper bound of affordable and politically viable sequestration potential around 5 gt per year for growing forests (some of which is making up for remaining deforestation emissions).[86]

Of course, the same pressures on land availability that cause deforestation also make growing forests hard in some regions. Efforts to grow new forests and to prevent current forests from being cut down will probably go hand-in-hand, as part of a single initiative in each country, with nonprofits partnering with government bodies to regulate large swaths of land. Coastal wetlands and peatlands might hold the most easily tapped potential for sequestration from ecosystem restoration in the near future.[87]

SOIL MANAGEMENT

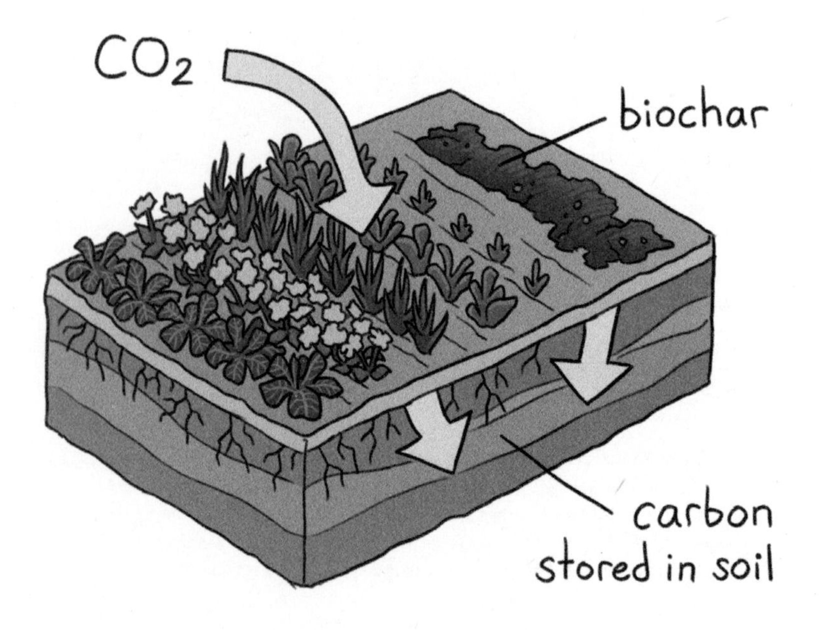

Similarly to forest soils, agricultural soils store large amounts of CO_2. These soils, too, are often disturbed, which releases CO_2. Better management can not only stop those emissions, but also sequester more CO_2 from the atmosphere in the soils. Various agricultural practices, such as using cover crops and rotating crops and livestock, can build up more carbon content in soils long-term. Spreading biochar, a concentrated form of carbon made by heating food or farm waste, on soils can store some of that carbon in the ground while making soil nutrients more accessible for plants as organic matter coats the porous surface of biochar.[88]

These cheap forms of sequestration have even more direct co-benefits than those associated with growing forests. Better soil management can reduce farmers' costs and improve their yields and therefore profits. This is one category of "clean technology options" that is, in fact, already cheaper. The problem is that most farmers are used to different practices and skeptical about changing their business models, especially for small or uncertain benefits. So, massive education and partnerships with farmers around the world are needed to get these practices adopted. Policies in relevant countries could incentivize adoption of more sustainable soil management practices, and could support the efforts of organizations conducting outreach to farmers to explain the practices and their benefits.

More research is also needed on soil management techniques—in Chapter 8 we noted the conflicting studies about whether tilling soils is good or bad in terms of emissions. Rotating crops is generally seen as good, and cover crops (which are then tilled into the soil) can certainly add carbon to the ground, but exactly how much potential soil sequestration holds and what practices are best pursued in what locations is relatively uncertain.[89]

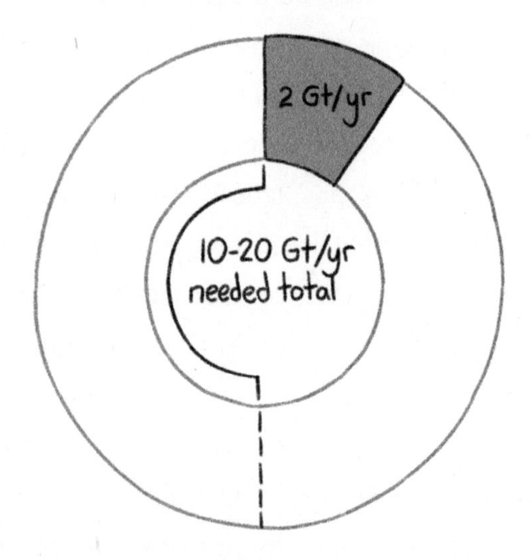

The upper limit of sequestration per year from soil management is determined by the amount of agricultural land in the world, and an estimate of what percent of farmers might be convinced to switch their practices within our thirty-year timeframe. Estimates put the feasible and affordable limit at around 2 gt per year.[90]

BIOENERGY WITH CARBON CAPTURE AND STORAGE (BECCS)

One proposed method of carbon sequestration is growing plants—trees, grasses, anything—to burn in power plants for electricity or in industrial plants for heat (or gasify to turn into other products), and to use normal CCS (carbon capture and sequestration from power plant or factory exhaust) technology to permanently store the resulting CO_2 emissions. The idea is that the whole process is carbon negative: plants capture CO_2 from the air as they grow, and then when we burn them we store the CO_2 underground. CO_2 makes up a much higher percent of power plant or factory exhausts than it does the atmosphere, so capturing it at the factory is generally much cheaper than capturing it from the atmosphere. With BECCS, companies don't have to go through the currently expensive step of filtering CO_2 directly out of the air—the plants do that for them—and they get the co-benefit of a potentially cheap form of energy for power plants and industrial sites. As noted in previous chapters, one major drawback with any plan that relies on biomass is that when companies grow the plants to begin with, they're likely taking away land area that otherwise could have grown food crops (or if the plants in question are trees, they could be sustainably harvesting them and constructing buildings with them, which sequesters the carbon in them and displaces steel and concrete). Anything that contributes to continued pressures toward deforestation is causing emissions, so depending on how BECCS plants are harvested, they could be causing more emissions than they subtract. If agricultural land is converted to grow biomass for BECCS, the total need for food crops remains the same and the agriculture industry will have to convert

Exhaust without CO_2
Wood
Crop waste
Impermeable clay layer
CO_2

land somewhere else in the world to make up the difference. If demand for plants—especially trees—to burn grows too large, this may directly become a form of deforestation.[91]

BECCS needs to be approached with caution, but it may have a role to play. Any biomass that can be sustainably harvested from forests (therefore giving an economic reason to keep forests as forests rather than to clear-cut them for pastureland or such) and burned with CCS would presumably be carbon negative (assuming only clean energy is used in its transportation and processing). We will have to create a strong monitoring and certification system to make sure that when someone buys "sustainably harvested" wood to burn, it really is sustainably harvested. Also, waste biomass—corn stalks, for instance—that is generated as part of growing food could be excellent feedstocks to create biofuels that are then burned with CCS and have genuinely negative emissions. Some of these wastes already contribute a couple percent of global emissions as they decompose or are otherwise processed (the amount of CO_2 they release is the same as what they captured when growing, but depending on how they decompose or are burned, they often release methane or nitrous oxide as well),[92] so using CCS would at least sequester those emissions while displacing some other fuel. However, a serious concern with BECCS is that—like any solution that relies on CCS equipment—it will only happen in areas where policy to mandate it is politically viable. Using

CCS is always more expensive than running the same process without the extra CCS equipment, and while sustainable biofuels are lower carbon than fossil fuels—some might even be considered carbon neutral—they're not a form of sequestration in themselves. They require the CCS component to make the process carbon negative.

SEQUESTRATION POTENTIAL OF BIOENERGY WITH CARBON CAPTURE & STORAGE

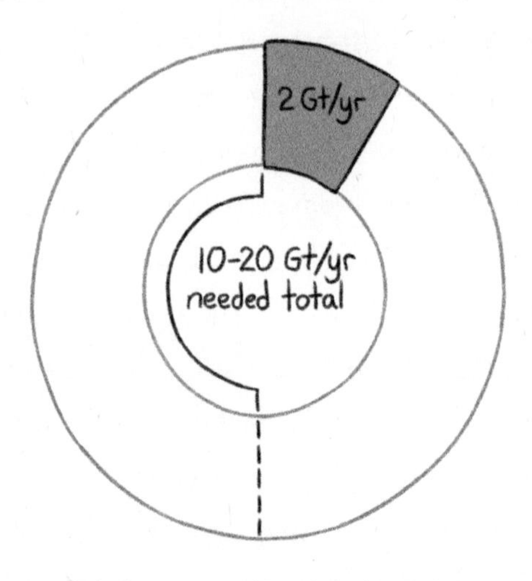

Estimates of sequestration potential are highly uncertain for BECCS because they depend on assumptions about indirect impacts on deforestation and other emissions. Some estimates say 0.5–5 gigatons per year, some closer to ten; two is a good estimate for what we can be reasonably confident won't increase deforestation emissions, and will be affordable.[93]

DIRECT AIR CAPTURE (DAC)

The most obvious and direct form of sequestration, direct air capture (DAC), generally involves large arrays of fans that blow air past catalysts that bind to CO_2, which is then separated in concentrated form by either heating and/or wetting the material.[94]

Some proposals involve a variant of this process called direct ocean capture (DOC). Because CO_2 is 140 times more concentrated in the ocean than in the atmosphere, it should require less effort to filter it out of seawater than air—if equipment can be designed to handle the salt and other materials dissolved in seawater that would flow through it. Because CO_2 mixes between the atmosphere and ocean to balance the concentration of CO_2 in each (currently with a net flow from atmosphere to ocean), climate scientists hypothesize that if we removed CO_2 from seawater, more CO_2 would dissolve into the ocean from the air to re-establish that equilibrium. This would remove a certain amount of CO_2 from the atmosphere.[95]

Once CO_2 is filtered from the air or ocean, something has to be done with it. The simplest option is to pump it deep underground where it is sealed in geologic formations from which it is unlikely to escape, at least for a much longer time than the relevant timeframe for solving climate change. This is called geologic storage. Other options exist, though, which may have more economic benefits: for instance, captured CO_2 can be

Air
Ethylene
(carbon stored in plastic, instead of making plastics from oil)
Electricity
Hydrogen
CO_2
Synthesis factory
CO_2
Seawater
Seawater without CO_2
Impermeable clay layer
CO_2

turned into ethylene, the feedstock chemical for most plastics. Turning CO_2 into solid commodities can lock the carbon up for many decades or centuries.

Both DAC and DOC are currently expensive. One company has successfully demonstrated multiple DAC plants at a cost around $500–600 per metric ton of CO_2 captured. Costs for forest and soil sequestration range from slightly negative (soil sequestration saves farmers money or makes them more money than it costs) to perhaps a bit under $100 per ton. DAC (or DOC) must come down in cost to something around or below $100 per ton in order to be an economically "reasonable" means of sequestration—a target that companies are optimistic about hitting. Currently three companies are working on DAC technology, and various academic or military labs are working on both DAC and DOC.[96] Direct capture plants could also serve a role as a flexible load to balance clean electricity generation, similar to low-capital-cost fuel synthesis plants.

Geologic storage is only starting to be demonstrated, but the science behind it is fairly well understood. CO_2 will eventually be pumped into saline formations under several layers of impermeable rock. Right now, we can start to sequester some CO_2 by pumping it into mostly depleted oil fields, which is already done with CO_2 from elsewhere underground to increase the underground gas pressure and push out the last dregs of oil. It sounds counterintuitive, but if done with atmospheric CO_2 it can sequester more CO_2 than the amount released when the recovered oil is burned (sometimes that will happen because of the amount of CO_2 required to achieve the desired pressure, but even if not, extra CO_2 could be added beyond what is necessary to push out the oil, in response to policy incentives to do so).[97] This "Enhanced Oil Recovery" is obviously a niche market and a poor long-term option, but can provide an immediate way to start capturing and sequestering CO_2 to prove various pieces of the technology.

Some labs are also developing processes to synthesize permanent materials based on captured CO_2. By making plastics with captured CO_2 instead of oil, the world would further reduce the market for oil, and sequester CO_2 in the plastic materials we use all around us. If all the ethylene currently used in the world were made from captured CO_2, it would only sequester the equivalent of 1% of current emissions, but that's a large enough sliver to be worth pursuing.[98] Other materials, such as carbon fiber for building with, could potentially also be synthesized from captured CO_2. And of course, the same air or ocean capture technology is a precursor to creating synthetic carbon-neutral (but not carbon-negative) drop-in fuels.

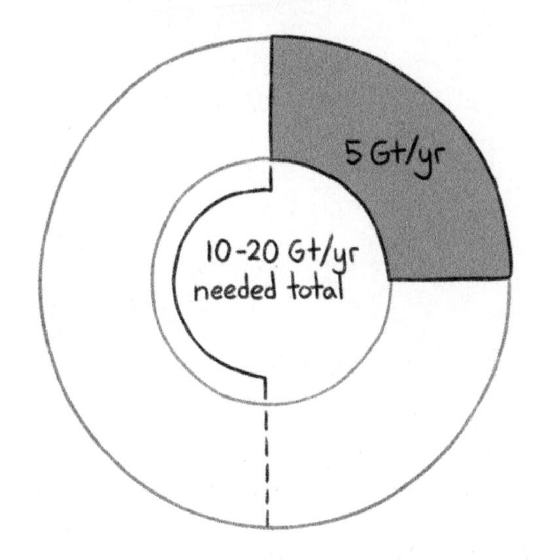

Theoretically, DAC/DOC sequestration has limits far higher than the amount of CO_2 we expect will ever need to be sequestered. The question is how fast costs can come down and how much governments will be willing to pay. Five gigatons per year is based on projections of likely cost improvements.[99]

ENHANCED WEATHERING/OCEAN ALKALINITY ENHANCEMENT

Naturally, over time CO_2 binds with certain silicate rocks and turns into minerals, which eventually end up in water systems and the ocean as calcium carbonate. Speeding up this process by crushing these rocks and spreading them out (to increase the surface area available for atmospheric CO_2 to bind to) could sequester a portion of the CO_2 needed. The crushed rocks could

be spread on farmland, where they would bring added benefits to soil pH and perhaps reduce the need for fertilizer. They could also be spread anywhere else appropriate on land, or on the ocean itself. The minerals that make their way into the ocean will slightly balance the effects of ocean acidification (which harms fishing and coral reefs). Enhanced weathering has yet to be researched at large scale, but it is expected to be fairly cheap, though not nearly as cheap as forest and soil sequestration methods. The main problem enhanced weathering may face is that mining and processing rock at the scale needed to achieve a significant amount of sequestration could be more energy intensive than it is worth (compared, for instance, to

DAC if DAC becomes much cheaper) or simply impractical in terms of the amount of infrastructure needed to gather, grind, and distribute the crushed rock. Farmland, where extensive infrastructure already exists to process large amounts of fertilizer, lime, and other commodities, might be the best site to spread the crushed rock if enough can be created.[100]

We need more research to determine if there are potential negative side effects, as well as benefits, to ecosystems. Other research must explore the impacts that such a massive scale of mining might have on humans, determining if any people would be harmed by particulate matter drifting from where crushed rock is processed or spread. But enhanced weathering is currently expected to be fairly benign.

One benefit of enhanced weathering is that it would add alkaline (basic) materials to the ocean and adjust its pH. This is essentially an antidote to ocean acidification: as CO_2 has been released into the atmosphere, more than half of it has mixed into the ocean, making the ocean more acidic (lower pH). This damages all sorts of marine ecosystems, such as coral reefs, quite apart from the impacts of climate change. Adding alkaline materials to the ocean would increase its pH and make up for some of the acidification.

Therefore, a related proposal is ocean alkalinity enhancement, which would consist of dumping other materials such as calcium oxide (lime) that aren't themselves binding with carbon, but will do so over time in the ocean. This increases the amount of CO_2 that the ocean will pull from the atmosphere each year, while still shifting the overall pH balance slightly away from the dangerous levels of acidity that have come from CO_2 emissions.[101]

Similarly to direct ocean capture proposals, more research may be needed to understand how the mixing of CO_2 changes when CO_2 is removed from, or alkaline materials are added to, the ocean. The results may also depend on how fast the world continues to add CO_2 emissions to the atmosphere.

Lime-based alkalinity enhancement runs into the same potential issues as enhanced weathering: depending on how many gigatons of sequestration we have to rely on it for, it may not be logistically or economically practical to process that volume of minerals. Lime, as we noted in the context of cement production, currently causes significant CO_2 emissions when it is refined from limestone, so lime used for ocean alkalinity enhancement would have to be produced with CCS. As is the case with other kinds of enhanced weathering, research should be done to discover any negative side effects that alkalinity enhancement might cause, though it is expected to be generally beneficial as it makes up for the acidification that is currently damaging ocean ecosystems.[102]

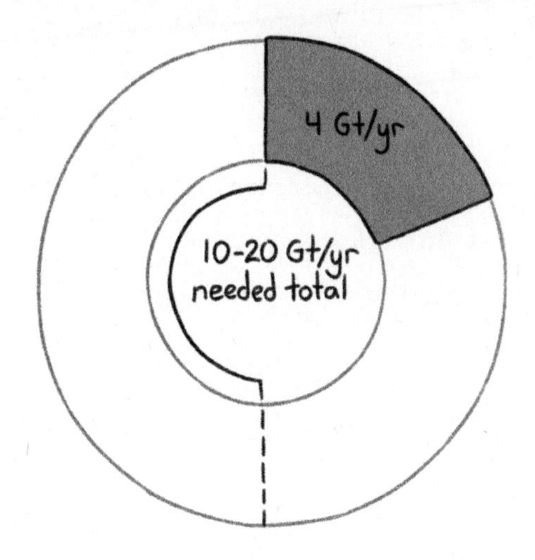

Given the constraints of scale and land, estimates put the upper limit of enhanced weathering and ocean alkalinity enhancement sequestration at around 4 gt per year.[103]

OCEAN FERTILIZATION/OCEAN AFFORESTATION

One sequestration method that could be cheap but needs more research is dumping iron into sections of the ocean to create bursts of growth in algae, which capture CO_2 as they grow and then sequester it as they die and sink to the ocean floor. In areas of the ocean with little or no existing ecosystem, this could be a good idea. Research needs to be done to test concerns that sudden growths and deaths of algae could create harmful side effects.[104]

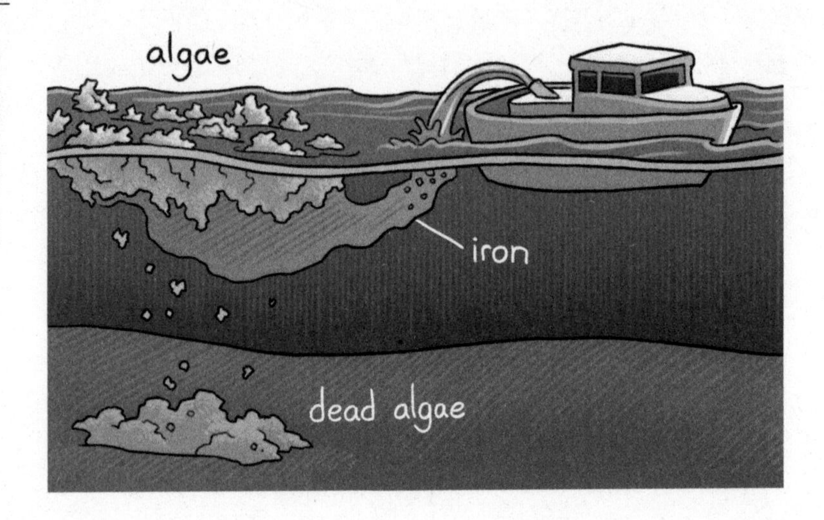

If research supports its use, ocean fertilization could be a cheap and simple method to implement, but one with no co-benefits. A variant could be to start scaling up ocean-based agriculture, combining seaweed growing and harvesting with fishing, perhaps algae growth, and maybe other crops as well. There are certainly large swaths of ocean with virtually no ecosystems present, and to the extent that food crops could be harvested at sea, this could reduce pressures toward deforestation on land.

One proposal lays out a vision for ocean afforestation, meaning growing forests of seaweed in the ocean just as we would grow forests of trees on land. The ocean has dramatically more area in which to do so, though the logistics of planting and harvesting are more complex. This particular

proposal includes letting some of the seaweed decompose and sequester its carbon as it falls to the ocean floor, but mostly it calls for harvesting the seaweed and using anaerobic digesters to turn it into methane and CO_2. The methane would be a presumably carbon-neutral biomass fuel, and could displace fossil methane (though in distributing it, there might still be some methane leak emissions). The CO_2 would be sequestered either underground or in pipes or similar at the bottom of the ocean.[105]

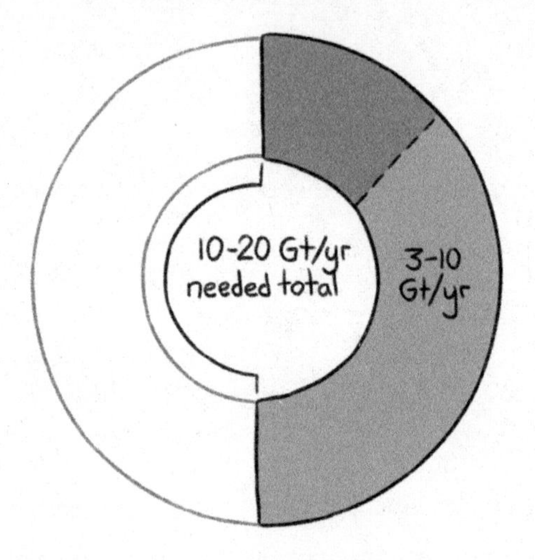

Sequestration potential for ocean fertilization/afforestation is a rough estimate, because not much research has been done on this method's effectiveness or acceptableness. Some researchers, based on possible negative side effects and the massive scale needed to add up to significant amounts of sequestration, suggest fertilization is not worth pursuing at all. On the other hand, if ocean afforestation ended up being viable at massive scale, its potential sequestration could reach over fifteen gigatons per year. However, it has only been prominently studied by one research team. Three to ten gigatons per year is a good range assuming that ocean afforestation won't reach gigantic scale by 2050.[106]

PROSPECTS FOR SEQUESTRATION

Growing forests and managing soils to store carbon stand out as the cheapest methods of sequestration. But they can only sequester perhaps a total of 5–10 gigatons of CO_2 per year. We will likely need significantly more sequestration, which is where the more expensive but less limited options come in.

Enhanced weathering, ocean fertilization/afforestation, and bioenergy with CCS could contribute an additional few gigatons, depending on what research into their potential scale-up and side effects reveals. Beyond that, direct capture and geologic storage (or use in permanent goods such as plastic or building materials) will be needed. This "technological" realm of direct air or ocean capture and conversion to products or geologic storage is in the early stages of active development.

Three companies and a few more academic labs currently claim to be able to capture CO_2 from the air, but none are currently near the "reasonable" cost of $100 per ton of CO_2.[107] Government-led coordination to convene and share ideas among labs working to improve this technology would accelerate the cost reduction. Funding for such labs and startups, and perhaps a centralized facility available for companies to bring their equipment to test for free and get an objective rating of its merits, would also be helpful. Policy incentives to subsidize capture and sequestration, or to penalize industrial carbon emissions and thereby give companies an incentive to capture and sequester emissions from their plants or from the atmosphere, could speed along the scale-up of these technologies as well.

A 2019 report from the energy analysis firm Rhodium Group suggested that government purchasing power could also drive the buildout of DAC equipment, by procuring carbon-neutral fuels made with DAC CO_2, thereby guaranteeing a market for the capture equipment.[108] A mandate, as part of a clean fuels standard or broader clean energy standard, could also do the same. The report also suggested creating a national carbon capture agency to fund, oversee, and carry out capture and sequestration projects.

By 2050, government bodies will have to foot the bill for each year's direct sequestration (except the bits that go into plastics and other valuable products) enough to make up for any remaining emissions and get into negative emissions. If that price tag is a couple hundred billion dollars per year worldwide (the lowest that can reasonably be expected by 2050), it might happen. If it is a couple trillion, probably not.

EFFORTS FOR SEQUESTRATION

The sequestration pillar demands government funding to support academic and nonprofit research into nature-based sequestration methods, and to drive technological improvements to lower direct air capture costs. It also relies on the global outreach initiative discussed in Chapter 8 for implementation of soil and forest management practices with farmers, nonprofits, and governments around the world to accelerate the use of nature-based sequestration methods. The technological sequestration methods will be energy intensive, so cheap, clean, readily available electricity is also a prerequisite.

By 2050, total costs for sequestration need to be low enough that government bodies are willing to pay for those with outright costs. However, perhaps more significant than the cost per ton of sequestration will be how many tons need to be sequestered every year.

$$\text{cost of sequestration per year} = \text{cost per ton} \times \text{number of tons required per year}$$

That number will depend largely on how many tons of remaining emissions need to be made up for each year. Reducing emissions is almost always cheaper than sequestering the equivalent amount of CO_2 from the atmosphere, so getting emissions as close to zero as possible—carrying out Pillars 1–4—is another prerequisite for the success of the 5th and final pillar of sequestration.

PART 3

HOW TO GET TO WORK

10

BEYOND THE FIVE PILLARS

NON-INNOVATION POLICIES

The five pillars constitute the minimum of what is "enough" to add up to a 100% solution. Without all five being implemented fully in some way, we will not reach negative emissions by 2050. However, there are many other steps that fall outside the bounds of the minimum work required to implement the pillars, but might either support their implementation or supplement that work with other benefits.

In political work, advocacy asks must usually be focused and simple to have the best chance of success. Often like-minded-organizations have to agree on a unified set of asks to get them enacted. Any efforts that pretend to be moving toward full solutions but in fact cannot add up will actively inhibit climate change work. However, any efforts that acknowledge the role of their ask(s) in the context of a 100% solution and strategically choose asks and advocacy tactics to align with other efforts can be extremely beneficial in advancing climate change work as a whole.

For instance, efficiency mandates, as discussed in Chapter 3, can never add up to zero emissions, but they can make the total size or cost of the energy system smaller, thereby easing the transition. Mandates for efficient truck engines and heating systems, or incentives to better insulate buildings or use public transit (or walk or bike) instead of driving can all reduce the total demand for electricity, synthesized fuels, and even new equipment itself. This could end up playing a significant role in making the required scale-up of manufacturing of new equipment, forms of fuel, or electricity generation achievable. If the world has to build 100 PWh of new electricity generation by 2050 instead of 150, it doesn't change the viability of the overall work needed to get there, but it may well make the difference between getting there in 2048 or in 2055.

A possible government initiative related to efficiency

policies is deployment of public infrastructure.[1] Building better train systems, mass transit in cities, or public charging stations for electric cars can all reduce total transportation emissions by making it easier for people to choose non-emitting options. The main consideration when comparing these initiatives to direct innovation or tech scale-up funding is whether the cost of the public infrastructure in question is reasonable enough and the expected emissions reductions or system cost reductions are significant enough to be worth it. If deploying new train systems that few people are going to ride takes away money from innovation efforts on organic solar cells or scale-up support for electric cars, then it will be doing more harm than good. If a new train system can be designed and built for a reasonable cost in a way that dramatically reduces the need to develop synthesized diesel for trains or trucks, or if the money is "extra" beyond what would otherwise be spent on innovation work, then it may bring significant net benefits to society.

Policies can also steer the economy to one clean technology over another, either to make regulation more politically viable, or to bring other long-term benefits. Perhaps one electricity generation option would be in itself the fastest way to decarbonize an area, but another would give better reliability or resilience against storms. A technology-specific subsidy or mandate could ensure the adoption of the desired technology.

That doesn't make the 100% solution more or less likely, but it changes what that area looks like once the solution is achieved. The key with these sorts of measures is to only preference technologies that can actually serve whatever portion of energy demand a region is relying on them for.

Some policies can also ensure a more equitable transition, bringing serious benefits to people, even if they aren't strictly necessary to achieving a 100% solution.

Such equality-driven policies include investments in developing countries by industrialized countries, siting regulations to ensure new heavy infrastructure isn't concentrated in low-income communities, job training programs to make sure everyone can participate in the new industries that are

> **Justice-oriented policies may be essential to the politics and implementation, if not the physical engineering, of any number of solutions in various countries.**

going to be created, and early retirement payments for fossil fuel workers.[2] In addition to furthering worthy causes beyond strictly solving climate change, these policies might turn out to be essential political measures: by including them as part of a comprehensive proposal, political leaders may find more support for carrying out their plans as a whole. Justice-oriented policies may be essential to the politics and implementation, if not the physical engineering, of any number of solutions in various countries.

ADAPTATION

There is a category of work almost entirely distinct from implementing the five pillars, but important as well: adapting to climate change impacts. Even if the world successfully reaches negative emissions by 2050, we will still see significant changes in weather patterns and sea levels by that point, and beyond.[3] Even as atmospheric CO_2 levels come down and temperatures slowly drop to historic averages, we can expect the middle and late part of this century to be marked by significant, if not massive, climate change impacts.

Policies to incentivize construction away from the coast can reduce flood and storm risk and associated damage costs. Public infrastructure deployment in the form of seawalls and other safeguards can protect communities from the increased threats. Plans to manage wildfires effectively when they are more common and more intense will be necessary. And political efforts to ensure that people displaced by climate change impacts will be welcomed in other parts of their home country or in other countries around the world will be one of the defining challenges of 21st-century international relations and domestic policy.

Adaptation measures can save and improve lives, so long as they come alongside a full implementation of the five pillars so that climate change impacts eventually lessen. If greenhouse gas emissions are not brought to zero, in a matter of decades the impacts will become too immense to adapt to.

GEOENGINEERING

One realm of adaptation worth exploration is geoengineering—blocking a small fraction of incoming sunlight with reflective particles dispersed in the high atmosphere by planes, thereby slightly slowing warming. Current technology could be assembled together into a workable system for dispersing sulfur particles in the stratosphere in a matter of months.[4]

Because these particles drift toward the Earth's poles and in about two years return to the ground, they need to be constantly replenished. But the amount of sulfur needed to slightly slow warming, and the associated costs, are very small compared with other climate change solution and adaptation measures.

If geoengineering were relied on as a major part of preventing impacts, without a 100% solution achieving negative emissions, the amounts and costs would eventually become unmanageable. But if a 100% solution is implemented and the world has certainty that temperature averages will slowly come back down to historic levels, geoengineering can shave off the peak of the temperature increases we would otherwise see.[5]

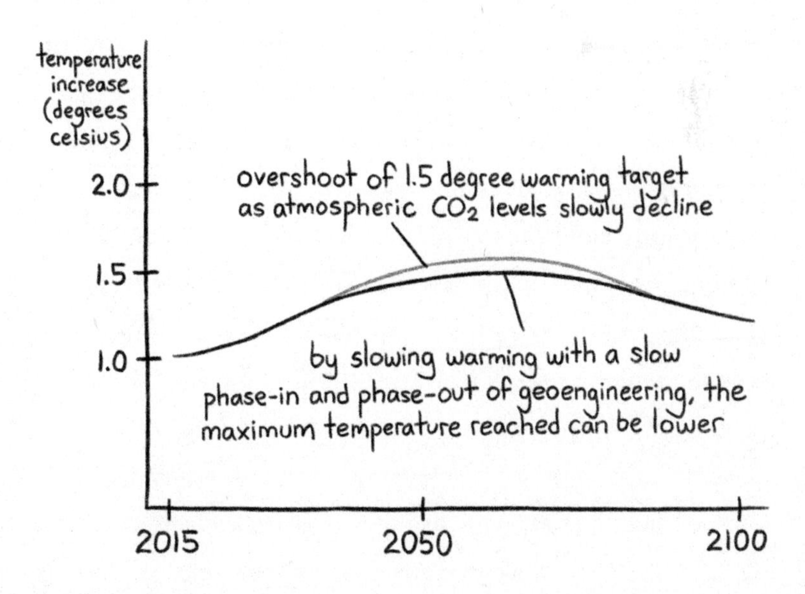

Essentially, geoengineering can improve the stability of society during this century, while negative emissions ensure stability for future centuries.

Partly because people misunderstand geoengineering's potential and think it could be relied on as part of a solution (as opposed to merely a temporary adaptation measure), the idea has gotten a lot of flack from advocates and politicians. Certainly no conversation about geoengineering should imply that it can substitute for eliminating emissions. But when considered as part of adaptation to the immediate impacts of climate change, geoengineering is probably one of the cheapest and widest-reaching measures to take.

Geoengineering with reflective particles does have side effects, such as acid rain from the sulfur particles, and the possibility of uneven impacts on temperature across the globe. This makes the international politics extremely hard: If Russia is slightly harmed but all of Africa and the Middle East benefit significantly, how can countries negotiate to try it in the first place? This is especially complicated given the relatively low costs of geoengineering, which mean virtually any large country or coalition of small countries could implement it without needing the support of the full global community.

Generally, the few scientists actively studying geoengineering favor an even dispersal of particles over the whole globe, which they expect will produce the most uniform cooling effects.[6] It is possible to disperse particles only over a certain area. The particles would eventually spread, but effects could remain concentrated in one region. This, scientists worry, could lead to unintended side effects impacting other regions by shifting the comparative temperatures and weather patterns. More research and modeling should be done to determine whether the full-globe approach is indeed the only sensible path.

There are other suggested approaches for bouncing away a portion of incoming light, including developing new particles that could replace sulfur to avoid the acid rain side effects and possibly also reduce cost.[7] Researchers have also suggested spraying salty seawater from boats on the ocean into clouds, which brightens the clouds and makes them more reflective.[8] This is a purely local form of geoengineering and only works where you can navigate boats to execute it. Some scientists worry that this option is either technically impractical or too expensive, or that it would run into the same hypothesized problems of skewing weather patterns and causing side effects as would the uneven dispersal of particles in the stratosphere. Others see it as a much more manageable option than the high-atmosphere sulfur particles. Cloud brightening could be employed in certain crucial areas (along with studies to prevent throwing off weather patterns in other regions). For example, one proposal suggested using cloud brightening only for the hottest weeks of the year to protect the Australian coral reefs from further bleaching.[9]

Another idea could be to slow warming locally over the Arctic. Because the North Pole is currently warming at a faster rate than the rest of the globe, the unevenness of slowing warming there without slowing warming across the rest of the earth might actually be a benefit rather than a further disruptor. This local geoengineering could be achieved with either cloud brightening (at least for Arctic areas near enough to the ocean) or sulfur particles (which drift toward the poles, so if spread around the Arctic Circle should stay over the pole before falling to earth again). The main benefit might be slowing the release of methane from Arctic permafrost as it melts, which would slow the total accumulation of greenhouse gases in the atmosphere and thereby slow overall warming.

All experts agree that geoengineering should be slowly phased in and out to avoid drastic impacts associated with the rate (rather than the amount) of temperature change. This will require careful control and international agreement, alongside continuous research, modeling, testing, and monitoring. Political struggles complicate geoengineering, but it holds the potential to improve lives this century as we take on the work of the five pillars.

Another approach to geoengineering focuses on glaciers: specifically, engineering them so they don't slide into the ocean and raise sea levels. A crucial site for this technology would be the West Antarctic Ice Sheet, parts of which some scientists think are already on an irreversible track to slide into the ocean and raise sea levels by as much as several meters.[10]

The ice shelf slides when its bottom layer melts, creating a slick surface that allows the ice shelf to slip off the solid land of Antarctica and into the ocean. The process will happen over hundreds of years, but the eventual magnitude of sea-level rise from this and a couple other melting glaciers makes it perhaps the most significant climate impact in terms of harm to humans. One proposal suggests removing the liquid water from under the ice shelf, so as to slow this slippage. Another proposes building artificial islands along the coast to simply block the path of the ice shelf so it can't move further into the ocean.[11]

Artificial islands could possibly prevent ice shelves from sliding into the ocean.

These are longer-term measures, to prevent the most catastrophic impacts of climate change, which should begin being pursued now but which are much less urgent than implementing all five pillars to achieve a 100% solution of negative emissions by 2050.

REVOLUTIONARY SEQUESTRATION

In addition to adapting to the impacts that are unavoidable even as temperatures slowly return to historic average levels, the rate of negative emissions each year will determine how quickly temperatures return to those levels. The more emissions are eliminated and the more research and innovation are done to make the six sequestration methods discussed viable to push to their full potential, the faster temperatures will reach historic averages.

But if we want to see that "return to normal" during the later part of this century—which would be the best way to minimize human suffering and loss of other species and ecosystems—we will need more than twenty gigatons of CO_2 sequestered every year. By a decade or so beyond 2050, we would need to be sequestering closer to fifty gigatons every year, about the amount of total greenhouse gases we're currently emitting each year.[12]

To make this larger amount of sequestration viable, more revolutionary technologies might need to be invented. Ideas have been floated for various methods, such as genetically engineering redwood trees to grow much faster, or using genetically engineered bacteria to turn CO_2 into other chemicals that could be easily stored underground or at the bottom of the ocean.[13] And perhaps with more research, some of the near-future methods discussed in Chapter 9 (especially the ocean-based methods, which have the advantage of cheap natural processes and a gigantic area in which to work) would turn out to have greater potential than expected.

11

POLITICAL LEADERSHIP

For many years, climate change sat on the political back burner. Certain politicians in various countries tried to take significant steps, and in a few rare instances succeeded, but polls showed consistently low interest in the issue among voters.[14] Between partisan polarization, efforts misspent on convincing denialists that climate change is real, and a lack of strength among the small activist movements that existed, nothing major has been accomplished up to this point. The world is still emitting greenhouse gases—faster than ever.

However, this apathy is on its way out. As climate change impacts have begun to present themselves in everyday life—even if they are only a hint of what's to come—and as United Nations and other scientific reports have issued more dire warnings and urgent calls for action, climate change has risen suddenly to be a top-tier issue in many countries.[15]

In both the United States and Europe, youth movements have brought new levels of attention to the issue and the need for action. High schooler Greta Thunberg, who started a "school strike" outside the Swedish parliament, galvanized students around the world to protest and draw political leaders' attention. The Sunrise Movement put the phrase "Green New Deal" on the map and forced a bolder level of policy proposals from US Congress members and presidential candidates.

As an example of this paradigm shift, in the 2016 US presidential election, almost no debate questions were asked about climate change. In the 2020 election, most candidates put forward reasonably ambitious climate change plans early in the primary. Also, China and India have started taking more significant action on clean energy in order to reduce extreme levels of air pollution in major cities.

All of this has created a moment in which the required scale of political action, which even three years ago might have been next to impossible, now seems within reach. A 100% solution will depend on whether political leaders in the next couple years harness this enthusiasm and make this moment the turning point.

> **To have any chance of a 100% solution adding up by 2050, the world needs serious, bold leadership.**

As we've seen, a 100% solution cannot be achieved by adding up incremental efficiency tactics, or by waiting for widespread behavior changes or for the natural pace of technology innovation to catch up. Carrying out all five pillars fully in thirty years requires a scale and pace of action that has only ever been seen during the World Wars.

For this reason, it is not enough for countries to elect politicians who are on the "right side" of the issue. It is not enough for legislators to simply vote the right way or for presidents and prime ministers to simply nudge along the incremental progress that has started. To have any chance of a 100% solution adding up by 2050, the world needs serious, bold leadership.

THE UNITED STATES

The United States is the wealthiest large country and has a strong track record of driving innovation. If it can muster the level of bold leadership needed, it has the most power to create and carry out the moonshot-type efforts we need. In the United States, most of the relevant leadership needs to come from the next president. The president sets the tone for national policy action, has far more convening power than any other individual politician, and is responsible for international partnerships and negotiation. As referenced in Chapter 4, the manufacturing boom leading into World War II and the Apollo Program were both possible only because of ambitious visions of bold presidents.

The next US president is the best-positioned person in the

entire world to start the moonshot initiatives we need. The president can create new teams, coordinate the work of various federal agencies, shift massive amounts of money to where funding is most needed, and empower all kinds of staff and partners to carry out each required piece of the effort. The president can ask Congress for the necessary level of funding, and JFK proved that it's possible to get it. In the last couple decades, Congress and the president have often been at odds, but visionary projects catering to national strengths such as American innovation and industry can create a common goal for Congress and the president.

The president also has to sell the idea to the public. One serious obstacle that prevented President Obama from accomplishing as much as he wanted on climate change regulations was that it was hard, especially given the rise of the Tea Party (which blasted any idea Obama proposed, even conservative ones), to message on climate change action in a way that got the public on board. If an initiative—even one that the president has complete theoretical power to launch—is perceived to be controversial, or wasteful, or partisan, it can be impossible to carry out for fear of the lost political capital that would damage future effectiveness of that president's administration. Luckily, innovation-focused solutions are already popular, so the public only needs to be sold on the scale, not the concept. The president's efforts will mostly need to be focused on explaining the full scope of actions needed so people understand why hundreds of billions of dollars have to be invested and what benefits they can expect in return.

The president is also responsible for international negotiation, which could be key to securing the level of funding needed for moonshot initiatives. International coordination is also crucial for preventing deforestation—especially in rainforest countries—and promoting policies that encourage sustainable farming practices. Various international agreements, trade deals, aid packages, and other initiatives could enable or speed the adoption of certain or all clean technologies in various countries. Such efforts could also ensure a more just transition to negative emissions by supporting communities that are hardest hit by early climate change impacts and communities that are least able to afford changes in infrastructure or lifestyle.

Finally, the president has a vast amount of convening power beyond work with Congress and foreign governments. The US president can convene corporate leaders and nonprofits. White House initiatives can bring together industries to establish best practices and push companies toward meaningful commitments related to their energy use, supply chain, or messaging. Some companies and individual investors might prove useful partners in funding pieces of the moonshot initiatives. Others might be able to scale up technologies quickly into the marketplace. Others might have inroads in certain countries, with foreign companies, customers, or governments, to supplement the diplomatic clout the president holds. Ultimately, it falls to companies to sell everyone the clean equipment that gets developed, so the more they can be involved and inspired toward bold actions, the faster the transition will happen.

CHINA

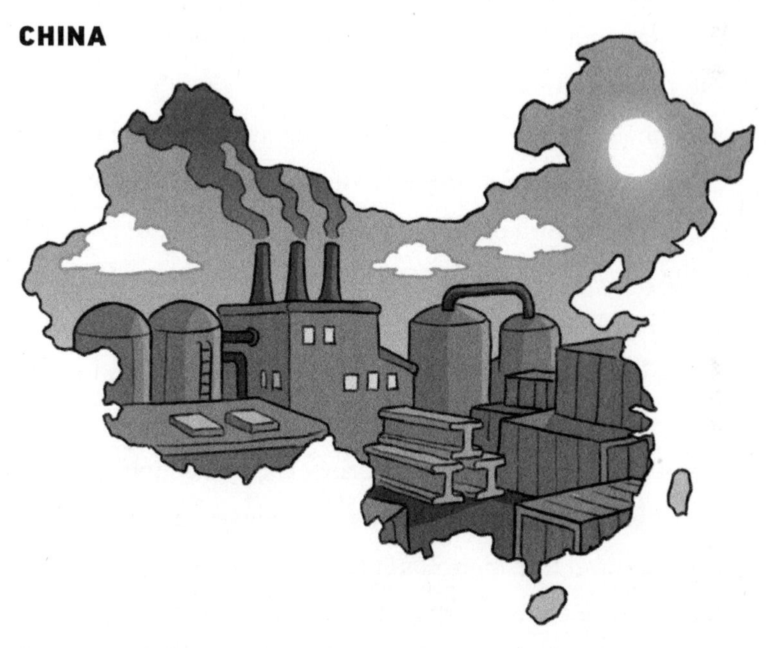

Because of the country's large economy, technological innovation, and centralized government, the president of China holds the power to single-handedly initiate most of the required work. The Chinese president is not nearly as limited by fear of political messaging failures or loss of power as the US president. China has high-tech research institutions that could carry out a lot of the needed basic engineering. China is of course known for its massive scale industry and people power, which could rapidly scale up standardized designs of new equipment. And China's economy as a whole is large enough that it could fund moonshot-type efforts on the required scale. China is probably the only country that could carry out the necessary projects single-handedly if the United States fails to lead.

One main difference between China and the United States is that US activism more directly affects the agenda of the president and what kind of proposals thought leaders put forward. In China, there's less room for activism to influence national policy initiatives.

Luckily, the goals of the Chinese government's worldwide development projects could align with climate change–solving projects. For example, outreach to farmers, prevention of deforestation, and deployment of new grids, power plants, industrial facilities, and transportation infrastructure could all be part of the vision of China's Belt and Road Initiative.[16]

EUROPE, AUSTRALIA, AND OTHER INDUSTRIALIZED COUNTRIES

No country beyond the United States and China has a large enough economy to carry out all five pillars fully and single-handedly, but in partnership with other countries, or by pressuring larger countries or international companies, any number of industrialized countries might be able to ensure that all the needed work gets done to add up to a 100% solution.

In most industrialized countries besides the United States, national governance happens in a parliamentary system. This can have significant advantages over the US system for implementing bold and rapid energy, industry, or agriculture shifts. Of course, parliamentary leaders are limited by all the same political communications challenges as the US president—with even more consequence, because a significant failure in messaging could not only limit their political clout during the rest of their term, but end their term far earlier than expected through a vote of no confidence or new national elections.

On the flip side, as long as it can successfully sell the idea, the party in power can sometimes unilaterally institute a program it sees as most urgent. For example, in British Columbia in 2008, the conservative party decided to implement a carbon tax. It was fiercely criticized, but the party played the messaging game well, won all the subsequent elections, and lost only a decade later to a coalition that wanted to increase that carbon tax even further.[17] That kind of initiative can work very quickly when it requires only the leadership of one party. In systems where a president or prime minister guides the policy of the coalition in power, they can be the bold leaders that make such action happen.

Many-party systems can create more room for ideas to be embraced across party lines than in the US two-party system, which promotes an assumption that whatever one party proposes the other will be against. Coalition governments mean that a leader must sell an idea not only to their own party, but to other parties in the coalition. This can slow down initiatives, but it can also make parties more open to one another's ideas, which is crucial to transcending the polarization that has halted US progress.

The problem with most industrialized parliamentary

countries is that they have far smaller populations, economies, and national budgets than the United States or China. A moonshot effort that the US president could fund with no one else's approval might require ten or twenty European countries pooling their money to achieve.

It would take a coalition of these countries to fund all the necessary work in each of the five pillars. This would require significant international cooperation and trust in whatever central entity is being funded from many countries' budgets. Smaller groups of countries or individual countries can of course push forward certain components on their own that we know will be necessary in a 100% solution, but they should also remain focused on ways they can draw in more partners or pressure larger players to take the necessary action. Smaller steps can make it more likely that the larger steps become politically or economically viable, but the largest steps will still only happen if *someone* is thinking comprehensively.

Such smaller initial steps could be pursued by various industrialized countries alongside efforts to move larger countries or industries to the full-scale solutions needed. For example, a parliamentary leader could sell the idea to their party and the public that they're going to convert the country entirely to electric cars—quieter roads, cheaper maintenance, healthier air, and helping to scale up a needed climate change–solving technology all in one. The same could be done with air-source heat pumps, industrial processes such as new cement production processes, or farming rules.

Basic research can also play a role. Australia, for example, has conducted some of the most promising research related to hydrogen and ammonia synthesis and conversion.[18] Government investments in such efforts, and partnerships with public and private institutions to scale up technologies that emerge, could make new options available for the rest of the world.

Any industrialized country can also exercise international leadership in the same way the US president can, working with developing countries to prevent deforestation, promote sustainable farm and forest practices, and protect communities who might get left behind amid either the impacts of or solutions to climate change. European countries in particular could also use their international clout to pressure the United States to take bolder action if US leaders don't step up. Various midsize countries could do the same with China through trade agreements, mutual commitments, and corporate or nonprofit partnerships.

Any number of countries could also convene companies and create initiatives to shift industries toward sustainable practices and adoption of clean equipment. In fact, some of the most significant global brands—including fossil fuel companies—are owned by or closely tied to individual governments in Europe and Asia, and their wealth and global reach could be deployed as a tool in shifting energy systems worldwide.

OTHER DEVELOPING COUNTRIES

Much of the work of a 100% solution has to be done by industrialized countries, which have the money to fund innovation projects, scale up new technologies, and pay for the adoption of clean options that are still slightly more expensive than fossil fuels. Among developing countries, only China, which is ahead of most in terms of economic growth, has enough money to guarantee on its own that all five pillars are carried out. Other developing countries can still partner with each other or with certain industrialized countries or companies to scale up specific pieces of technology. Again, that won't lead directly to a 100% solution, but it can make it more likely that larger economies will take the needed actions to add up to 100%. Many developing countries can also take leadership roles on agriculture practices, which are generally cheap (often even saving money or increasing farmers' total profits) and simply require convening, outreach, education, and commitment to adopt.

Finally, depending on how cheap various technologies get, some developing countries may simply not be able to eliminate as large a portion of their emissions as industrialized countries can. Industrialized countries, including the United States and smaller ones, could give funding to enable deployment of marginally more expensive technologies in developing countries. Or industrialized countries could subsidize the manufacturing of such technologies, as China has done for years with solar panels.

ACTIVISTS

For those of us not in the top positions of leadership in our countries, our goal is to influence those leaders to take the necessary actions across the full scope of solutions needed. Whether we are think-tank analysts in Brazil, grassroots activists in the United States, or staff of a membership nonprofit that focuses on climate change worldwide, *we* must keep a comprehensive mindset and marshal our efforts with coordination and communication.

For activists in the United States, the 2020 election presents a huge opportunity to affect national priorities and the scale and tone of climate change discourse. Already, the work of activist movements, especially the Sunrise Movement, in the last two years has heightened the attention given to climate change and the scale of solutions being proposed.[19] Most candidates' specific proposals still aren't quite at the level of achieving a 100% solution, though—or at least, they don't specifically lay out enough of what they intend to do for us to know whether they would really meet all the criteria of this framework.

Whether in the US 2020 elections or in upcoming elections for Congress, or parliament or presidency in any country, campaigns are a time when politicians are out talking to and listening to voters, and forming their agenda for the coming term. By showing up at campaign events and asking public or private questions, by writing e-mails and letters about policy, by getting press attention that forces candidates to respond to certain policy proposals, and by working to elect the candidates that seem most committed, we can shape the climate change agenda for the coming years.

In any political effort, communication is key. Activists' asks must be unified, simple, and focused. Too many conflicting, confusing, or extraneous demands by various organizations can muddle the discourse, distract policymakers from the most crucial proposals, or make activist movements seem less powerful than they could be. Partly, this is what this book is meant to help with: providing a set of criteria for what is the minimum of "enough" on climate change action. That can form the frame of reference for our asks and guide our critiques of proposals that politicians put forward.

Recent climate change movements have learned from what worked and didn't in previous efforts. For example, up until this decade, climate change was usually discussed as an "environmental" issue, conjuring ideas about saving distant species, highlighting the intangible nature of the problem, and making the issue more polarizing (at least in US politics). Now, more and more people are talking about climate change as a human issue, one that could dramatically disrupt our economy, safety, and public health, not to mention destabilize global politics with shifts in migration patterns and resource availability.

Some people have even moved on to a frame of communications that's often best: not talking about climate change at all. The phrase "climate change" is still controversial to some people. But the *solutions* to climate change are generally not. Everyone supports cleaner energy, healthier air, more local choice and control, vibrant natural spaces, and less disease. Everyone wants cheaper energy and agrees that more people in the world deserve access to modern energy infrastructure. People love the idea of innovation, and of new, better technologies creating new jobs across our countries in labs, factories, and construction sites.

Many of us who started our climate change advocacy in the realm of carbon pricing learned this quickly: carbon pricing would never pass as a "solution to climate change." That sounds like we're making some kind of collective sacrifice to save the polar bears and perhaps the abstract future generations of humans. But carbon pricing brings significant benefits to the economy—making business more efficient, increasing overall incomes, *and* reducing pollution. Many carbon pricing advocacy groups, most notably Citizens' Climate Lobby in the United States, have formed successful bipartisan and nonpartisan coalitions because one doesn't even have to believe in climate change science to support carbon pricing: it's good for the economy and public health, and so are all the other components of a 100% solution to climate change.[20]

Innovation is popular, and often focusing on the innovation aspects of a solution can garner interest and support from folks beyond the usual supporters. Relatedly, presenting solutions with a comprehensive framing can earn more credibility, generate more excitement around the strong vision, and reduce controversy by breaking people out of their default modes of thinking with the creativity of the proposal. Tying various pieces together into a comprehensive proposal can get more support for individual pieces that might on their own be controversial, but which make logical sense in the context of the larger vision that people can buy into. And of course, comprehensive messages and demands make it more likely that comprehensive action will be taken, which is the only way we can expect a 100% solution by 2050.

> **Solutions-oriented rhetoric is more unifying, exciting, forward-looking, and practical.**

It's important that we talk about solutions rather than problems. Sometimes a reminder of the dire impacts climate change will bring can move already converted activists to new levels of commitment. But often it contributes more to defeatism or denialism. Most of the time, solutions-oriented rhetoric is more unifying, exciting, forward-looking, and practical. That means it's more focused, and more likely to spur policy action.

COMPANIES

A final set of actors in the leadership picture are companies, which are often either not considered or demonized when it comes to political involvement on climate change. The fact remains, though, that most people interact far more regularly with the corporate sector than with government institutions. People's views of the world and ideas about many issues, political or personal, can be shaped by advertisements, company brand messaging, announcements of new initiatives businesses are taking, and news about corporate happenings.

Companies, then, have a significant amount of messaging power. They often use this reactively—for example, making a new commitment on an issue they think people care about in the hope that it will draw more customer loyalty to their brand. But those messages also flow in the other direction, affecting people's perception of not only the company but the issue in question.

Companies that want to make a meaningful difference on climate change can start by not "greenwashing." This practice of emphasizing how "environmentally friendly" a company is because of some measure that saves a bit of water—or how their bottles use a little less plastic, or how their materials are recyclable, or more and more often how they are moving toward "100% renewable electricity" (without mentioning the "net," because of course they still buy fossil fuel electricity at night)—reinforces the dangerous idea that efficiency tactics could add up to a full solution to climate change.

Companies that are serious about climate change solutions should instead shift their messaging to reflect the comprehensive picture of action needed. A good example is how Google has bought enough renewable electricity to equal the total amount of power its data centers use, but has also talked openly about its "net" 100% renewable power and discussed how hard it is to get the timing of renewable generation and data center electricity demand to match.[21]

Companies can also start making some of the specific pieces

happen. It's possible that a broad coalition of companies working on the various pieces and donating to nonprofits that carry out the farm and forest work could in fact add up to a 100% solution. More likely, companies can enable key pieces of the solution to move forward and by setting such an example, and using messaging to make it clear they understand the larger picture as well, make it more likely that national governments will follow.

Some pieces of the solution that companies are well equipped to push forward are emerging technologies that have not yet been scaled up. Companies can use their purchasing—and messaging—power to speed their scale-up. For example, companies that distribute products around a country or the world could pioneer electric (or hydrogen/ammonia) truck technology and build out the charging (or refueling) infrastructure necessary. Not only would they likely save money on transport costs over ten or twenty years, but they would create the first round of infrastructure that would allow smaller companies and the transportation industry as a whole to shift to carbon-free options.

Similarly, many large companies have extensive supply chains. This puts them in a position to monitor and influence practices among the producers or wholesalers they purchase from. Fair trade and organic certifications are the most prominent examples of this, but there are plenty of other labels and criteria that companies can insist their suppliers meet.

Especially for companies that deal in either wood or agricultural products, requiring certain standards from their suppliers can have a serious impact on global sustainability. For example, in 2006 several major food companies jointly created a "soybean moratorium" to protect the Amazon Rainforest in Brazil. These companies agreed not to purchase soybeans grown on land that was deforested after the moratorium began, and alongside other efforts, this led to significant decreases in deforestation in Brazil.[22] Such voluntary business commitments can be spurred by grassroots activism—consumer pressure on companies is often as effective as voter pressure on political candidates. We underuse the activist tactic of pressuring businesses, but it should feel natural because businesses are the entities that have to physically implement most of the energy and agriculture shifts needed in the five pillars.

New business coalitions could be created to label and require "deforestation-free" products, which could go a long way to helping developing countries keep in check both legal and illegal clear-cutting, as the market dries up for products produced in that way. At the same time, this could help support farmers across the world who are using sustainable practices.

Many other supply chain issues could be tackled by the world's larger companies. Electric cars, emissions-free building heating, even ships and airplanes, could be scaled up through the purchasing power of major international corporations. A company could require that all of its office buildings convert to be net-zero energy. A company could ensure that all fuels it uses are synthesized carbon-neutral ones. Consumer movements could pressure and encourage whole industries to make these shifts with more and more ease as innovation lowers the cost of various clean technology options.

It is unlikely that company efforts alone will add up to anywhere close to a 100% solution. But they are useful tools alongside other activism to achieve 100%.

PURSUING ALL OPTIONS

When considering what companies and what small- or medium-sized countries can do, we must remember that the best hope of achieving a 100% solution to climate change by 2050 is for the United States (or maybe China) to take a bold leadership role and devote hundreds of billions of dollars per year to funding moonshot projects and related efforts.

As activists and thought leaders, we have to stay focused on that goal. To prevent the most disastrous effects of climate change, we should aim to set *all* the needed work in motion in the next couple years. However, we shouldn't adopt an "all at once *or bust*" approach. Hopefully we will make the full-scale projects happen, but if we don't immediately, the smaller efforts we can simultaneously pursue—for example, to directly start scaling up some of the most crucial pieces of technology—can be a way of "hedging our bets." On the chance that US or Chinese leadership totally fails to materialize in the next couple years, the leadership of other countries and companies, spurred by activists, can still achieve some chunks of emissions reductions that buy us slight bits of time for other solutions to roll out, or that directly pressure US or Chinese leaders to step up soon after.

The transition time is thirty years, and the sooner we take every step the easier, cheaper, and more popular the total transition will be. But as long as we ensure that steps add up to negative emissions by 2050, we will still achieve the minimum 100% solution to climate change.

12

KEY PRESCRIPTIONS

Building on the previous chapter's general advice for various sorts of countries, activists, and companies, in this chapter we'll lay out more specific and short-term prescriptions. Generally, this book has aimed to present a logically tight framework which acknowledges various possible options and the nuances of each. At the same time, some readers are probably looking for a prescriptive list of things we should work on. Below is such a list—a starting point consisting of the most immediate individual projects to be pursued.

These initiatives and policies would not alone add up to a 100% solution, but most of them are essential components of one. Many of them represent the largest chunks of emissions reductions we could achieve in the short term, thus buying time for other technologies to be readied and deployed, and decreasing the eventual need for sequestration.

NUCLEAR SCALE-UP

As discussed in Chapter 5, the most concentrated form of clean energy, and therefore the one that can scale up fastest, is nuclear. The main prospect for getting nuclear plants to the point that their capital costs are lower than coal or methane plants is to mass-manufacture them in shipyards with one (or a few) standardized design(s). The best shipyards for this purpose are currently in South Korea, though there are likely candidates in Japan as well. To add a large amount of electricity generation by 2030, many more shipyards will have to be built quickly. That's a task any industrialized country could undertake.

Similarly, the supply chain of steel, low-enriched uranium, turbines, and workers will have to be scaled up rapidly. Again, any country (or individual state or province in some cases) can bite off a chunk of the supply chain and start scaling up enrichment facilities or open a new institution to start training the bigger workforce we will need.

However, perhaps most important is getting the first batch of these standardized plants built. Unlike smaller pieces of technology, there's a lot of financial risk in building a multi-billion-dollar power plant.

Even if the whole project might make a lot of money, traditional investment isn't geared toward accelerating the demonstration phase and moving to the first commercial plant. Countries that are too small to finance the entire scale-up project may still be able to propel the project by becoming the first customers of these standardized plants. If the United Kingdom, for example, ordered ten standardized plants, and signed a power purchase agreement to buy the electricity at a rate equivalent to its current fossil fuel electricity costs, it could guarantee that there is a market for these products and thus spur investors to fund the manufacturing of the first batch of plants.

The physical success of a new battery or solar cell chemistry could be proven by hiring a roll-to-roll processing facility to make one from materials a research lab provides. Proving the physical success of a new nuclear design requires building one full plant, even if it's a small one.[23]

The standardized design in question could be a simplification of an existing design, such as the South Korean reactor that has already been exported successfully. Or it could be a new design that is quickly brought through the demonstration phase, as various startups are working on. The plants could be deployed on floating platforms, or barged to their end locations and installed on land as normal or on the seabed just next to shore.

This is a project that individual companies, or even certain individual people, could fund and carry out. Demonstrating a new design would cost about $1 billion and take perhaps five years if it were pursued vigorously. Adopting an existing design would cost very little. And each power plant would cost a couple billion dollars in the beginning, dropping as manufacturing is scaled up to about one billion dollars per one gigawatt plant—the mark at which they will be cost-competitive with coal and methane worldwide.[24]

HYDROGEN ELECTROLYSIS EQUIPMENT

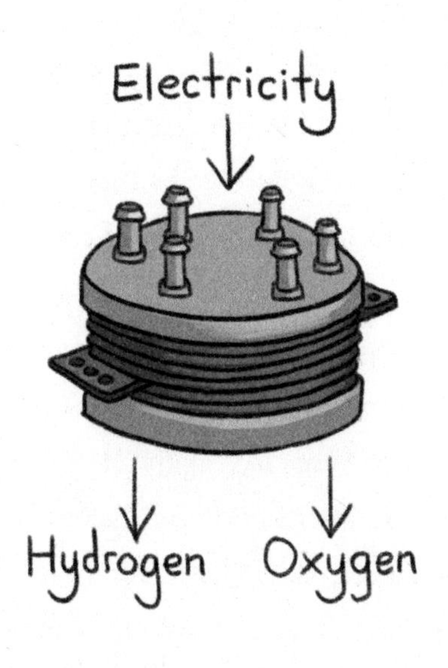

Another key technology that has to be developed is hydrogen electrolysis equipment with significantly lower capital costs. As discussed in Chapter 7, current electrolyzers cost so much up front that even with a couple decades of methane feedstock costs, methane-derived hydrogen is still cheaper than clean electrolysis-derived hydrogen. Yet hydrogen is an essential ingredient in ammonia production (for fertilizer or fuel) and for any drop-in fuel. As also noted, equipment with extremely low capital costs that could economically be run intermittently to balance renewables or nuclear on the grid would enable a faster and cheaper scale-up to 100% clean electricity generation.

For both those reasons, a top priority for lab and startup research funding should be revolutionary hydrogen electrolysis equipment. New catalysts, new membranes, new electrode materials, different designs of the systems, and cheaper manufacturing processes could all help achieve this goal. Various academic researchers are working on materials to serve in these roles, or on chemical pathways that could make hydrogen synthesis cheaper. Certain proposed but not commercialized hydrogen electrolysis methods, such as several that use high temperatures, could turn out to have lower capital costs as well.[25]

Large companies already build hydrogen electrolyzers at scale, but most are currently optimized for efficiency given the input cost of electricity. If electricity becomes much cheaper through the nuclear scale-up, or in certain locations with abundant solar or wind sites, hydrogen electrolysis plants might be able to be designed that are optimized for lowest possible capital cost rather than efficiency. Design and commercialization projects as part of moonshot efforts should focus on electrolysis plants that can achieve the lowest possible capital costs even at the expense of efficiency, because the up-front cost is usually the main factor in decisions about whether to build new infrastructure (whether or not it would save money over thirty or more years), and because clean electricity costs are likely to continue decreasing over the coming decades. This is the opposite of the mentality of academic research, which rewards efficiency improvements (the inventions that add to scientific knowledge) rather than capital cost improvements (the inventions that make for short-term commercial viability). Academic labs usually don't have many barriers to buying or creating high capital cost devices, either. This is another example of the need for coordination and focus driven by moonshot leaders to convene engineers who might not otherwise have incentives to work on commercial-focused rather than science-focused designs.

And as noted in Chapter 7, some proposals for clean hydrogen don't rely on electrolysis at all, but on converting biomass wastes or dedicated biomass crops into hydrogen fuel. The wastes in question are dispersed in small amounts across many farms, so aggregating them takes a significant amount of energy in itself—in some cases, more energy than one would get by turning that amount of waste into fuel. Making space for dedicated crops threatens to create more deforestation, but if pursued alongside serious reforestation and afforestation efforts and with the most high-yield crops available, the net impact could be quite positive. CCS with methane-derived hydrogen can also be used, but like all CCS will only happen where policy mandates it. That could be an easy intermediate, short-term policy step certain countries can take.

Hydrogen itself can be used as a fuel, but many researchers and startups are developing processes to turn it into carbon fuels or other products. These conversions can happen relatively affordably already—not cheaply enough to compete with fossil fuels, but enough to make mandates or subsidies viable and to possibly outcompete remaining fossil fuels after electrification. Hydrogen, ammonia, and carbon fuels can all be interconverted relatively easily. The missing piece is an initial way to generate any one of them, and hydrogen is the most likely to become that first step (and most versatile to then be turned into the other products). If clean hydrogen can become cheaper than fossil-derived hydrogen, this will take care of a significant portion of the work needed in Pillar 3.

ELECTRIC CARS AND BATTERIES

This is a simple one. Electric cars are already cheaper to operate than gas cars over either's lifetime.[26] All electric cars need is to be scaled up in manufacturing so their capital costs come down to the same level as those of gas cars. A lot of the difference in cost is from the batteries in electric cars, so entities with research and testing capital should fund improvements in lithium-ion batteries, or inventions for post-lithium-ion batteries. Batteries are a good case for testing a wide range of possible materials to eventually find one or more that work well, even if that means many disappointments along the way. The basic science that drives battery degradation, charging effectiveness, and such isn't totally understood. But the outcomes that we care about are easy to measure. Rather than trying to figure out exactly what chemical side reactions are messing up the intended working of a given battery cell (the academic focus that develops new scientific knowledge), this is a case for large-number trial and error (the commercial focus that gets a product deployed sooner). Unlike nuclear, battery cells can be tested at small size to determine which exact configuration of materials might work. Then, once a design is proven, they can be scaled up with little risk.[27]

Aside from batteries, electric cars simply haven't been manufactured at anywhere near the scale of gas cars, so a simple increase in the order of magnitude of manufacturing, with improvements to efficiency in business practices and so forth, could bring the cost of electric cars down to reasonable levels. States and countries can either subsidize or mandate the adoption of electric cars, and they can also purchase them in bulk for public car fleets.

The same goes for trucks. Although there are fewer government-owned big trucks, Amazon, UPS, and FedEx could take serious initiative and build the network of charging stations and procure the first batches of electric trucks necessary to electrify that industry.

Hydrogen tanks and fuel cells can substitute for batteries in cars and trucks, and offer much longer ranges (per weight and per volume taken up by the "battery") and shorter refueling times. The rest of the car remains the same, running on electricity produced in the fuel cell. Hydrogen fuel cells are even more expensive than lithium-ion batteries right now, so similar materials and manufacturing innovations would be needed to bring down the capital costs of these cars.[28] And as battery-based electric cars need charging stations to be deployed in larger numbers, hydrogen tank and fuel cell–based electric cars would require hydrogen fueling stations. Japan and a few other countries are putting government and corporate efforts into this approach and their model could be applicable across the world.

AIR-SOURCE HEAT PUMPS

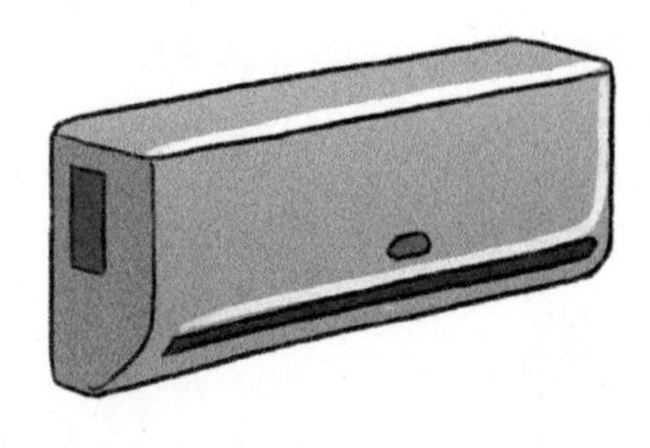

For most buildings, the best way to eliminate heating emissions will be to install air-source heat pumps, sometimes known as mini-splits, which run on electricity. They can also cool buildings, so their use makes particular sense for the swaths of India, China, and other warm regions of developing countries where demand for air conditioning is projected to skyrocket in the coming decades.[29]

Heat pumps are also in the category of technologies that can currently compete with fossil options, but aren't yet definitively cheaper. Research and testing funding can help. Scale-up can help, and this one is even easier than electric cars for governments to promote and support. Public initiatives could create programs to weatherize (retrofit for improved insulation and other efficiency) homes and commercial buildings, including installing heat pumps. Governments could bulk-purchase heat pumps to convert their own buildings. Policies could encourage or subsidize individual families or businesses to make the switch.

CEMENT AND STEEL

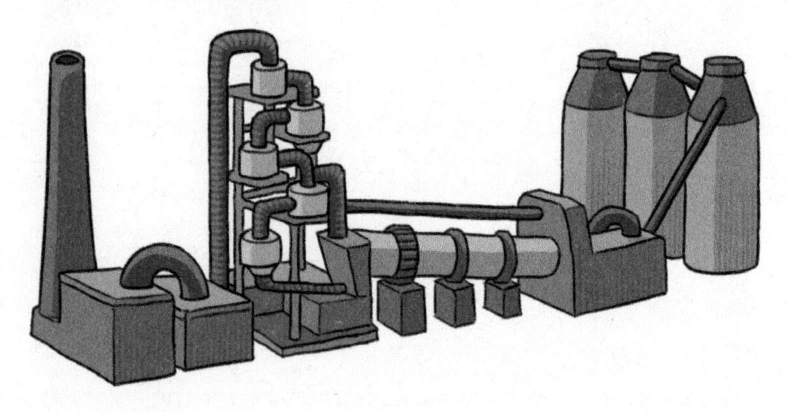

The largest industrial sources of CO_2 emissions are cement and steel production. As noted in Chapter 8, research and demonstration can be supported in order to develop new processes that reduce or eliminate emissions from these factories, and hydrogen can be used as a replacement fuel with some level of reconfiguration of the factory equipment. Getting these high-inertia industries to adopt these new practices and pieces of equipment might take several decades, though.

In the meantime, any country that can muster the political clout to do so should impose mandates or create incentives to get all steel and cement factories to use CCS equipment and sequester their emissions. This equipment will initially be expensive, but as it gets deployed at scale and industries learn best practices for its manufacture and use, it will drop slightly in cost so that at least the additional cost won't be wild. Between their fuel use and their process emissions, cement and steel combined contribute about 10% of total global greenhouse gas emissions, so mandating CCS for these factories would accomplish tremendous early reductions in emissions. A large number of these plants are in China, so this is a case where either Chinese presidential leadership or trade pressure from other countries—for example, policies prohibiting or taxing imports of steel or cement that don't have an objective certification of CCS use—will make a major difference.[30]

CCS MANDATES

More generally, in any region where it is politically possible to impose modest to significant extra costs for emissions reductions, one of the easiest ways to guarantee immediate emissions reductions would be to mandate (or subsidize) CCS for *all* factories and fossil fuel power plants. At a glance, this seems politically unlikely in most countries. But considering how fast political discourse has changed in the last two years, perhaps such significant mandates could be within reach. The cost for sequestering CO_2 from power plants is a little less than $100 per ton, which is not far from the eventual cost we will need governments to pay for atmospheric sequestration.[31] In fact, the more that is spent on CCS (less than $100 per ton) now, the less that will need to be spent on atmospheric sequestration (around or sometimes more than $100 per ton) later.

AGRICULTURE AND REFORESTATION

Another realm for immediate policy action on the non-energy side is growing or regrowing forests. National governments generally have significant power to designate large swaths of land as protected for certain uses. Simply leaving land alone for a few decades usually is enough for it to become a forest, especially if it had previously been a forest. With scientific management, new forests can be grown to be even healthier than they would be if left alone. Forests could also be optimized to capture CO_2 as fast as possible based on what species are growing in them. Sustainably managed forests can also be a source of wood for construction, which sequesters CO_2 into buildings themselves and also displaces cement and steel, which are far more emissions-intensive building materials.[32]

National leadership can force massive-scale reforestation and afforestation, but absent national leadership, communities in forest areas can also work to protect their lands, local governments can enforce laws against deforestation, and nonprofits can work to designate more and more sections of forest for protected management. As noted in Chapter 11, companies—mostly food companies—can also exert leadership by requiring their supply chains to be deforestation-free.

On farms, nonprofits have even more room to work. Organizations that train volunteers or staff to conduct outreach to farmers around the world can accelerate the adoption of sustainable farming practices. Such organizations can also partner with academic institutions to conduct continuous research and monitoring on various farms so that we understand better what practices to push for. And again, national and local governments could incentivize or in some cases require various crop rotation, soil management, cover crop, and fertilizer use practices. This work comes with the benefit of immediate cost-saving (or extra profit–making) opportunity.

CARBON PRICING

The most significant policy incentive to support the work of the five pillars is probably carbon pricing. The idea is simple: Fossil fuels and other sources of greenhouse gas emissions are bad and we want people and companies to use them less, so we'll make them more expensive. Usually, proposals charge the price at the first possible point (entry into the state for state-level policies, extraction from a well or mine or importation into the country for national-level policies) for efficiency. Most proposals include a specific use of the revenue generated from the price—often reinvesting it into the economy through either spending on clean energy, dividend checks (or tax rebates) to every resident of the state/country, or both. Perhaps most importantly, every proposal includes some sort of increase in the price over time, to give both individuals and businesses long-term certainty that clean options will be a good investment. The most prominent national US proposal, put forward by the single-issue carbon pricing advocacy group Citizens' Climate Lobby, is to start at $10 per metric ton of CO_2 that would be released from a given amount of fossil fuel, and to increase the price $10 every year, indefinitely.[33] The indefinite increase is the best way to guarantee that certainty—companies then know that paying up front for clean options *will* be a good investment over time.

Carbon pricing, as the vast majority of economists agree, is the most economically efficient way to reduce greenhouse gas emissions. The limitation of carbon pricing is that it takes a long time to ramp up to a price high enough to make every clean option cheaper than every fossil option. At the $10-per-year increase rate, it would take twenty years or more before virtually every clean option was definitively the cheapest. That doesn't leave enough time to then transition to electrified equipment and such before 2050. Still, along the way, the fact that everyone knows the price will eventually get there can create an incentive to transition—through early investments that companies know will pay off. The revenue raised can also be used to speed along various aspects of implementing the five pillars—for instance, supporting clean energy technology research or helping low-income families switch to electric home heating. And because it is the most economically efficient method of reducing emissions, carbon pricing can reduce the need for mandate-based policies to speed the implementation of the pillars.

Anywhere carbon pricing is politically viable, it should be implemented immediately, with the fastest possible ramp-up in price, and with that ramp-up continuing indefinitely to give economic certainty that the society will be transitioning entirely away from fossil fuels.

Carbon pricing can also be used to address some portion of non-energy emissions, if the price is applied to emissions-intensive goods such as cement and steel. When such products are produced at a factory or imported into a country, a price could be charged based on the non-energy emissions involved in making them. A well-designed pricing system might add a price to methane based on its fugitive emissions, and might also charge a price on livestock or other foods (at the point of production or importation) based on their average methane and deforestation impacts. Similarly, point-of-importation prices can account for the fossil fuel emissions embodied in goods manufactured overseas, and would be charged on imported goods coming from a country that does not have its own equivalent pricing system.

Carbon prices on imports, accounting for both energy and non-energy emissions, would incentivize other countries to impose their own pricing systems or to convert their industries to clean processes so their products don't face an added cost when imported into countries with robust carbon prices. This is especially important because while industrialized countries' economies have become lower in emissions intensity, much of their emissions have simply shifted to China and elsewhere—the places where most manufacturing now takes place. Industrialized countries can't claim to be carbon neutral if they rely on imports from places with carbon-intensive industry.

CONCLUSION

For those immersed in climate change research, energy systems engineering, and clean energy policy, most if not all the content of these five pillars may seem obvious. However, even if experts understand each of these pieces individually, it is rare for policy proposals or advocacy initiatives to tie more than a few of them together into a unified vision. It hasn't happened yet that a serious national political proposal in any country has laid out specific steps that add up to addressing 100% of the global problem of climate change by 2050.

Current youth movements, from high schoolers in Europe inspired by Greta Thunberg to Millennials in the United States participating in the grassroots Sunrise Movement, are putting climate change higher on the political agenda. This trend is likely to increase in the next few years as today's young generations become an increasing portion of voters, policy thinkers, and organization leaders.

But to achieve the goal laid out for us by climate scientists—negative emissions by 2050—government leaders will have to take steps at massive scale over the next thirty years. Given the timeframe for both developing and scaling up or rolling out technologies and practices, a lot of the relevant work has to happen in the next ten years.

With the urgency and the required scale in mind, activist movements must use these five pillars as a common frame of reference to judge proposals, make asks of political leaders, and ensure that steps eventually enacted add up to 100%. The timeframe is shorter than many people are considering. The scale of action needed is larger than most political leaders and organizations have acknowledged. And yet the solutions themselves—innovation, job creation through manufacturing scale-up and research, partnerships with companies, farmers, and governments around the world—are both politically popular in general and possible to carry out incredibly quickly under the right leadership.

As bolder climate change policy proposals develop, and (in the United States especially) as potential presidential initiatives take shape, it is crucial that activists and thought leaders consider proposals in the context of a 100% solution.

Two-thirds of emissions come from developing countries

that don't have the economic ability to adopt more expensive clean options like their industrialized counterparts can. So a solution that adds up fully by 2050 must make virtually every clean option cheaper than current systems. Industrialized countries must lead this effort—which will also benefit their own economies—by creating moonshot-style innovation projects in partnership with research labs, companies, and other countries. These projects can coordinate and fund research, development, testing, demonstration, and scale-up to ensure that each needed technology is cheap enough in time. The projects should include a global outreach initiative to shift farming practices, prevent deforestation, and start sequestering CO_2 in soils and forests. Policies including carbon pricing, mandates or subsidies for clean technologies that are still slightly more expensive than fossil options, and international pressure to speed each country in reducing emissions will play a role in guaranteeing negative emissions by 2050.

Through this work, a relatively small number of political leaders—the president of the United States or China, or a collection of European prime ministers, and the activists and thinkers who will influence them—can drive a thirty-year transition in the global energy and agriculture systems that achieves net-negative greenhouse gas emissions by 2050.

In the process, not only will leading countries boost their own economies, but they will guarantee greater energy access

We can leave behind the pollution and inadequate energy access of today...
H2 FUEL
...and build a clean and prosperous world for tomorrow.

in developing regions through lower costs. The health benefits from reduced pollution will be immense. Improved technologies, available to more people with fewer negative side effects, will rapidly improve quality of life and speed up economic development around the world. Solving climate change, by definition, will be one of the fastest improvements in economic well-being in human history.

That transformation is worth achieving, but it will not happen without intense levels of focused activism and leadership. It will not happen without people breaking away from the incremental mindsets that have defined climate change policy for decades. We must all keep in mind the comprehensive picture of what must happen to add up to a 100% solution. We must set aside cynicism and 10% thinking and acknowledge the need for global transformations. We must elect and support and pressure officials who will take the truly bold steps that can set the world on a path to negative emissions. The scale is enormous—new industries, massive public projects, trillions of dollars shifting toward cleaner investment—but the world we build will be more prosperous, just, and safe.

Now is the time to heed the call of our youngest leaders and act—and to do so with the comprehensive strategies that will make a 100% solution not only possible, but likely. Let's get to work.

ACKNOWLEDGMENTS

I am deeply grateful to the many people who made this book, and more generally the synthesis of this framework, possible. I'm sure I'll miss someone who offered essential advice or input, but I will attempt to name those who most actively shaped my work:

First, thanks to my parents, Andra Rose and Joshua Goldstein, who have both shifted into full-time climate change work at the prompting of their activist son. A special thanks to my dad, who has been my go-to person to discuss and flesh out my ideas—much of this framework developed through conversations while I paced around his house and we tried to wrap our minds around the enormous complexity that is the climate crisis.

Also an incredible thank you to my literary agent, Chris

Kepner, and editor, Ryan Harrington, both of whom took a chance on a new author with a youth voice, and who jumped into this project with their own youthful energy and passion. Their advice and determined professional efforts not only made the book happen, but turned it into the form it needed to be.

More thanks to my wife Sophia Normark, who learned with me and helped me talk through muddled ideas throughout this analysis. And to the rest of my family who contributed particular excellent ideas and advice.

A special thank you to Bill Budinger, who funded the creation of the book, making illustrations possible, and to my friend Violet Kitchen—whose art makes the book so much more useful and enjoyable, and who exhibited a depth of patience and skill in our numerous round-and-round e-mail exchanges on every single image.

This framework is synthesized from the expertise of many scholars, activists, entrepreneurs, and professors. Thank you to everyone who gave me input in person, by phone, or by e-mail, including: Keely Anson, Suzy Baker, Alex Barron, Paul Bauman, James Bradbury, Steve Brick, Jeff Brown, Trevor Brown, Jen Burney, Graciela Chichilnisky, Steven Chu, Armond Cohen, Steve Corneli, Steve Crolius, Steven Davis, Nick DeCristofaro, Stephanie DeLuca, Phil De Luna, Dave Devanney, Michael Ditmore, Maya Domeshek, Rick Duke,

Darcy DuMont, Richard Eidlin, Josh Freed, Julio Friedmann, Maria Gainer, Mengpin Ge, Deb Greenspan, Sarah Hunt, Rory Jacobson, Jesse Jenkins, Ryan Jones, Lars Jorgensen, David Keith, Mary Kraus, Brett Kugelmass, Klaus Lackner, Phil Larochelle, John Larsen, Joe Lassiter, Eli Lehrer, Kelly Levin, Nate Lewis, Heidi Lim, Renata Lippi, Lilly Lombard, Jessica Lovering, David McCabe, Tim McCall, Bill McKibben, Nathan Mueller, Deepika Nagabhushan, Dan Nocera, Emily Norton, Matthías Ólafsson, Christina Paxson, Miranda Peterson, Steve Pinker, Staffan Qvist, Peter Reinhardt, Phil Renfoth, Bill Ritter, Timmons Roberts, Catie Rutley, Jonathan Simonds, Varun Sivaram, Sarah Smith, Samantha Steiner, Tom Stokes, Amelia Stymacks, Jacob Teter, Laura Van Wie, Owen Woodcock, those who gave advice longer ago while I was state rep, and those whom I spoke with after the deadline to finalize the text of these acknowledgements.

ENDNOTES

PART ONE: SOLUTIONS MUST ADD UP

1 Valérie Masson-Delmotte et al., eds., *Summary for Policymakers: Global Warming of 1.5°C: An IPCC Special Report on the impacts of global warming of 1.5°C above pre-industrial levels and related global greenhouse gas emission pathways, in the context of strengthening the global response to the threat of climate change, sustainable development, and efforts to eradicate poverty* (Geneva, Switzerland: IPCC, 2018).

2 Ottmar Edenhofer et al., eds., *Climate Change 2014: Mitigation of Climate Change* (New York: Cambridge University Press, 2014).

3 William Nordhaus, *The Climate Casino: Risk, Uncertainty, and Economics for a Warming World* (New Haven: Yale University Press, 2013); and Jean-Daniel Collomb, "The Ideology of Climate Change Denial in the United States," *European Journal of American Studies* 9, no. 1 (2014): document 5, https://doi.org/10.4000/ejas.10305.

4 Masson-Delmotte et al., *Summary for Policymakers.*

5 Christina Nunez, "How Has Fracking Changed Our Future?," *National Geographic* (blog), accessed summer 2019, https://www.nationalgeographic.com/environment/energy/great-energy-challenge/big-energy-question/how-has-fracking-changed-our-future/.

6 Marc Gunther, "These cheap, clean stoves were supposed to save millions of lives. What happened?," *Washington Post*, October 29, 2015, https://www.washingtonpost.com/opinions/these-cheap-clean-stoves-were-supposed-to-save-millions-of-lives-what-happened/2015/10/29/c0b98f38-77fa-11e5-a958-d889faf561dc_story.html?noredirect=on.

7 Al Gore, *Our Choice: A Plan to Solve the Climate Crisis* (New York: Rodale, 2009).

8 James H. Williams et al., *Pathways to Deep Decarbonization in the United States* (San Francisco: Energy and Environmental Economics, 2014).

9 Mengpin Ge, Johannes Friedrich, and Thomas Damassa, "6 Graphs Explain the World's Top 10 Emitters," *World Resources Institute* (blog), November 25, 2014, https://www.wri.org/blog/2014/11/6-graphs-explain-world-s-top-10-emitters; and "C-ROADS World Climate," Climate Interactive, accessed summer 2019, https://croadsworldclimate.climateinteractive.org.

10 Hans Rosling, Anna Rosling Rönnlund, and Ola Rosling, *Factfulness: Ten Reasons We're Wrong About the World—and Why Things Are Better Than You Think* (New York: Flatiron, 2018).

11 "Data for High income, Middle income, Low income," The World Bank, accessed summer 2019, https://data.worldbank.org/?locations=XD-XP-XM; map created with MapChart.net.

12 "Birth rate, crude (per 1,000 people)," The World Bank, accessed summer 2019, https://data.worldbank.org/indicator/SP.DYN.CBRT.IN?locations=IN; "GDP (current US$)," The World Bank, accessed summer 2019, https://data.worldbank.org/indicator/NY.GDP.MKTP.CD?locations=IN; "Life expectancy at birth, total (years)," The World Bank, accessed summer 2019, https://data.worldbank.org/indicator/SP.DYN.LE00.IN?locations=IN; and "School enrollment, primary (% gross)," The World Bank, accessed summer 2019, https://data.worldbank.org/indicator/SE.PRM.ENRR?locations=IN.

13 "C-ROADS World Climate."

14 "C-ROADS World Climate."

15 "Air conditioning use emerges as one of the key drivers of global electricity-demand growth," *International Energy Agency* (blog), May 15, 2018, https://www.iea.org/newsroom/news/2018/may/air-conditioning-use-emerges-as-one-of-the-key-drivers-of-global-electricity-dema.html.

16 "C-ROADS World Climate."

17 "Renewable Energy Technologies: Cost Analysis Series—Hydropower" (working paper series, IRENA 1, no. 3/5, 2012).

18 "Brazil has the third-largest electricity sector in the Americas," U.S. Energy Information Agency, March 23, 2017, https://www.eia.gov/todayinenergy/detail.php?id=30472.

19 "Coal 2018: Analysis and forecasts to 2023," International Energy Agency, accessed summer 2019, https://www.iea.org/coal2018/.

20 Data sourced from many studies, some more recent than others. Most notably: Edenhofer, *Climate Change 2014;* and "Global Emissions," Center for Climate and Energy Solutions, accessed summer 2019, https://www.c2es.org/content/international-emissions/. Also checked against Steven Davis et al., "Net-zero emissions energy systems," *Science* 360, no. 6396 (June 29, 2018): eaas9793, https://doi.org/10.1126/science.aas9793; and checked against graphs personally sent by Prof. Steven J. Davis.

21 Pharoah Le Feuvre, "Commentary: Are aviation biofuels ready for take off?," *International Energy Agency,* March 18, 2019, https://www.iea.org/newsroom/news/2019/march/are-aviation-biofuels-ready-for-take-off.html; Umair Irfan, "Aircraft fuel is notoriously dirty. This airline is betting on clean electricity.," *Vox,* May 14, 2019, https://www.vox.com/2019/5/14/18535971/electric-airplane-aircraft-aviation-clean-energy; and Transport & Environment, *Roadmap to decarbonising European aviation* (Brussels: T&E, 2018).

22 Estimated remaining emissions are based on the analysis and relevant citations provided throughout this book.

23 In addition to citations in Ch. 9, see this report with a similar graph based on slightly different assumptions: Derek Martin et al., *Carbon Dioxide Removal Options: A Literature Review Identifying Carbon Removal Potentials and Costs* (Ann Arbor: University of Michigan, 2017).

24 "Surf N'Turf," Symbrosia, accessed summer 2019, https://www.symbrosiasolutions.com/solution; and "Agriculture's Impact," Greener Grazing, accessed summer 2019, https://www.greenergrazing.org/challenge.

25 Some examples: Climeworks (https://www.climeworks.com/); Carbon Engineering (https://carbonengineering.com/); Opus 12 (https://www.opus-12.com/); CERT (https://co2cert.com/); Carbon Recycling International (https://www.carbonrecycling.is/).

26 Doris Goodwin, "The Way We Won: America's Economic Breakthrough During World War II," *The American Prospect*, Fall 1992, https://prospect.org/article/way-we-won-americas-economic-breakthrough-during-world-war-ii. Image based on a government photo of a WWII airplane factory (see "File:Willow Run Factory.jpg" *Wikimedia Commons*, 2006, https://commons.wikimedia.org/wiki/File:Willow_Run_Factory.jpg.)

27 Franklin Delano Roosevelt, "Address Delivered by President Roosevelt to the Congress, May 16, 1940," May 16, 1940, https://www.mtholyoke.edu/acad/intrel/WorldWar2/fdr16.htm; and "300,000 Airplanes," *Air & Space*, May 2007, https://www.airspacemag.com/history-of-flight/300000-airplanes-17122703/.

28 "Moon Missions," NASA, accessed summer 2019, https://moon.nasa.gov/exploration/moon-missions/.

29 "The Sustainable Infrastructure Imperative," The New Climate Economy, accessed summer 2019, https://newclimateeconomy.report/2016/; and International Renewable Energy Agency, *Global Energy Transformation: A roadmap to 2050* (Abu Dhabi: IRENA, 2018).

30 International Renewable Energy Agency, *Global Energy Transformation*; and Dana Nuccitelli, "Climate change could cost the U.S. economy hundreds of billions a year by 2090," *Yale Climate Connections*, April 29, 2019, https://www.yaleclimateconnections.org/2019/04/climate-change-could-cost-u-s-economy-billions/.

31 Sean Pool and Jennifer Erickson, "The High Return on Investment for Publicly Funded Research," *Center for American Progress* (blog), December 10, 2012, https://www.americanprogress.org/issues/economy/reports/2012/12/10/47481/the-high-return-on-investment-for-publicly-funded-research/; "R&D pays: Economists suggest 20% return on public investment for research and innovation," *Science | Business*, June 27, 2017, https://sciencebusiness.net/news/80354/R%26D-pays%3A-Economists-suggest-20%25-return-on-public-investment-for-research-and-innovation; and Jonathan Gitlin, "Calculating the economic impact of the Human Genome Project," National Human Genome Research Institute, June 12, 2013, https://www.genome.gov/27544383/calculating-the-economic-impact-of-the-human-genome-project.

32 Steven Chu, "Climate Change and Needed Technical Solutions for a Sustainable Future" (Lecture, Amherst College, Amherst, MA, March 21, 2018).

33 An example of innovation's bipartisan popularity: Justin Worland, "President Trump Wants to Kill This Clean Energy Program Even Though It Has Bipartisan Support," *Time*, March 16, 2017, https://time.com/4703638/donald-trump-budget-energy-arpa-epa/.

PART TWO: WHAT HAS TO HAPPEN

1 International Energy Agency, *Energy Technology Perspectives 2017* (France: IEA, 2017); International Renewable Energy Agency, *Global Energy Transformation*; World Energy Council, *World Energy Scenarios: Composing energy futures to 2050* (London: WEC, 2013); and Williams et al., *Pathways to Deep Decarbonization.*

2 International Energy Agency, *Energy Technology Perspectives 2017.*

3 Proportions are illustrative estimates based on the analysis and sources referenced in this book.

4 International Energy Agency, *Energy Technology Perspectives 2017*; International Renewable Energy Agency, *Global Energy Transformation*; World Energy Council, *World Energy Scenarios*; and Williams et al., *Pathways to Deep Decarbonization.*

5 "Useful Life," National Renewable Energy Laboratory, accessed summer 2019, https://www.nrel.gov/analysis/tech-footprint.html; and "Energy Return on Investment," World Nuclear Association, updated November 2017, https://www.world-nuclear.org/information-library/energy-and-the-environment/energy-return-on-investment.aspx.

6 Varun Sivaram, *Taming the Sun* (Cambridge: MIT Press, 2018).

7 Sivaram, *Taming the Sun.*

8 "Featured Dashboard—Cost," International Renewable Energy Agency, accessed summer 2019, http://resourceirena.irena.org/gateway/dashboard/?topic=3&subTopic=33.

9 Thirty PWh max of wind from: Jami Hossain, *Wind Energy 2050* (Bonn, Germany: World Wind Energy Association, 2015).

10 Ryan Wiser et al., *Forecasting Wind Energy Costs & Cost Drivers* (IEA, 2016).

11 Graph courtesy of NorthBridge, based on Bruce Phillips, *Fully Decarbonizing the New England Electric System: Implications for New Reservoir Hydro* (Concord, MA: NorthBridge 2019). NorthBridge analysis based on ISO-NE data.

12 Current level from "Hydropower," International Energy Agency, accessed summer 2019, https://www.iea.org/topics/renewables/hydropower/; and technical potential from Patrick Moriarty and Damon Honnery, "What is the global potential for renewable energy?," *Renewable and Sustainable Energy Reviews* 16, no. 1 (2012): 244–252.

13 Jessica R. Lovering, Arthur Yip, and Ted Nordhaus, "Historical Construction Costs of Global Nuclear Power Reactors," *Energy Policy* 91 (April 2016): 371–382.

14 Junji Cao et al., "China-U.S. Cooperation to Advance Nuclear Power," *Science* 353, no. 6299 (August 5, 2016): 547–548.

15 World Health Organization, *Health Risk Assessment from the Nuclear Accident After the 2011 Great East Japan Earthquake and Tsunami, Based on a Preliminary Dose Estimation* (Geneva: WHO, 2013); and M. Ghiassi-nejad et al., "Very high background radiation areas of Ramsar, Iran: preliminary biological studies," *Health Physics* 82, no. 1 (2002): 87–93, https://doi.org/10.1097/00004032-200201000-00011.

16 The Chernobyl Forum (International Atomic Energy Agency et al.), *Chernobyl's Legacy: Health, Environmental, and Socio/Economic Impacts*, rev. ed. (Vienna: IAEA, 2006), 8.

17 Anil Markandya and Paul Wilkinson, "Electricity Generation and Health," *Lancet* 370, no. 9591 (September 13, 2007): 979–990.

18 James Conca, "Natural Gas and the New Deathprint for Energy," *Forbes*, January 25, 2018, https://www.forbes.com/sites/jamesconca/2018/01/25/natural-gas-and-the-new-deathprint-for-energy/#4eabe9955e19.

19 Gwyneth Cravens, *Power to Save the World: The Truth About Nuclear Energy* (New York: Alfred A. Knopf, 2007), 9.

20 Proportions derived from: Stephanie Weckend, Andreas Wade, and Garvin Heath, *End-of-Life Management: Solar Photovoltaic Panels* (IRENA and IEA-PVPS, 2016); "Size and weight of solar panels," EnergySage, October 1, 2018, https://news.energysage.com/average-solar-panel-size-weight/; Brian Wang, "Constructing a lot of nuclear power plants is not material constrained," Next Big Future, July 13, 2017, https://www.nextbigfuture.com/2007/07/constructing-lot-of-nuclear-power.html; "The Nuclear Fuel Cycle," World Nuclear Association, updated March 2017, https://www.world-nuclear.org/information-library/nuclear-fuel-cycle/introduction/nuclear-fuel-cycle-overview.aspx; Per F. Peterson, Haihua Zhao, and Robert Petroski, *Metal and Concrete Inputs for Several Nuclear Power Plants* (Berkeley: University of California, 2005); and "How much coal is required to run a 100-watt light bulb 24 hours a day for a year?," HowStuffWorks, accessed summer 2019, https://science.howstuffworks.com/environmental/energy/question481.htm.

21 Lovering et al., "Historical Construction Costs of Global Nuclear Power Reactors."

22 Gisela Grosch, "Generation IV Systems," Generation IV International Forum, September 23, 2013, https://www.gen-4.org/gif/jcms/c_59461/generation-iv-systems.

23 See: "Post-combustion coal fired power plant," British Geological Survey, accessed summer 2019, https://www.bgs.ac.uk/discoveringGeology/climateChange/CCS/PostCombustionCoalFiredPowerPlant.html; and Gary Shu et al., "Economics and Policies for Carbon Capture and Sequestration in the Western United States: A Marginal Cost Analysis of Potential Power Plant Deployment," Carbon Capture & Sequestration Technologies @ MIT, 2010, https://sequestration.mit.edu/research/westcarb.html.

24 Mike Mueller, "5 Things You Should Know About Geothermal Heat Pumps," U.S. Department of Energy, August 1, 2017, https://www.energy.gov/eere/articles/5-things-you-should-know-about-geothermal-heat-pumps.

25 For example: International Energy Agency, *Energy Technology Perspectives 2017*.

26 David Biello, "Deep Geothermal: The Untapped Renewable Energy Source," *Yale Environment 360*, October 23, 2008, https://e360.yale.edu/features/deep_geothermal_the_untapped_energy_source.

27 "Thorium," World Nuclear Association, updated February 2017, https://www.world-nuclear.org/information-library/current-and-future-generation/thorium.aspx.

28 "Wave and Tidal Energy," MIT Seed, accessed summer 2019, https://learning.media.mit.edu/seed/wave%20energy.html.

29 Ariel Cohen, "Is Fusion Power Within Our Grasp?," *Forbes*, January 14, 2019, https://www.forbes.com/sites/arielcohen/2019/01/14/is-fusion-power-within-our-grasp/#95b5d909bb4b; and "Nuclear Fusion Power," World Nuclear Association, updated July 2019, https://www.world-nuclear.org/information-library/current-and-future-generation/nuclear-fusion-power.aspx.

30 Matthew Davison et al., "Decentralised Energy Market for Implementation into the Intergrid Concept - Part 2: Integrated System," *7th International Conference on Renewable Energy Research and Applications* (October 2018): 3, https://doi.org/10.1109/ICRERA.2018.8566719.

31 Richard Moss, "New flow battery projected to cost 60% less than existing standard," *New Atlas*, December 22, 2015, https://newatlas.com/pnnl-low-cost-sustainable-flow-battery/41028/; and Zheng Li et al., "Air-breathing aqueous sulfur flow battery for ultralow cost electrical storage," *Joule* 1, no. 2 (2017): 306–327, https://doi.org/10.1016/j.joule.2017.08.007.

32 Akshat Rathi, "Stacking concrete blocks is a surprisingly efficient way to store energy," *Quartz*, August 18, 2018, https://qz.com/1355672/stacking-concrete-blocks-is-a-surprisingly-efficient-way-to-store-energy/.

33 Maya Wei-Haas, "Could Renewable Energy Be Stored in Balloons in the Ocean?," Smithsonian.com, January 6, 2016, https://www.smithsonianmag.com/innovation/could-renewable-energy-be-stored-balloons-ocean-180957603/.

34 "Air Source Heat Pumps," U.S. Department of Energy, accessed summer 2019, https://www.energy.gov/energysaver/heat-pump-systems/air-source-heat-pumps.

35 Nathaniel Bullard, "Electric Car Price Tag Shrinks Along With Battery Cost," *Bloomberg Opinion*, April 12, 2019, https://www.bloomberg.com/opinion/articles/2019-04-12/electric-vehicle-battery-shrinks-and-so-does-the-total-cost.

36 See ideas from Tony Seba.

37 John Larsen et al., *Capturing Leadership: Policies for the US to Advance Direct Air Capture Technology* (New York: Rhodium Group, 2019); and Ben Haley et al., *350 PPM Pathways for the United States* (San Francisco: Evolved Energy Research, 2019).

38 Chart adapted from Williams et al., *Pathways to Deep Decarbonization*. Used with permission.

39 "Completed Missions," SpaceX, accessed summer 2019, https://www.spacex.com/missions.

40 Transport & Environment, *Roadmap to decarbonising European aviation*.

41 Stephanie Searle and Chris Malins, "A reassessment of global bioenergy potential in 2050," GCB Bioenergy 7, no. 2 (March 2015): 328–336, https://doi.org/10.1111/gcbb.12141; and International Energy Agency, *Energy Technology Perspectives 2017*.

42 Grigorii Soloveichik, "Electrified future of aviation: batteries or fuel cells?," ARPA-E, March 13, 2018, https://arpa-e.energy.gov/sites/default/files/Grigorii-Soloveichik-Fast-Pitch-2018.pdf.

43 Akira Yamamoto, Masayoshi Kimoto, Yasushi Ozawa, and Saburo Hara, "Basic Co-Firing Characteristics of Ammonia with Pulverized Coal in a Single Burner Test Furnace," *15th Annual NH3 Fuel Conference* (October 31, 2018), https://nh3fuelassociation.org/2018/12/14/basic-co-firing-characteristics-of-ammonia-with-pulverized-coal-in-a-single-burner-test-furnace/.

44 Transport & Environment, *Roadmap to decarbonising European shipping.*

45 Davis et al., "Net-zero emissions energy systems"; and U.S. DRIVE Partnership, *Hydrogen Production Tech Team Roadmap* (U.S. DRIVE Partnership, 2017).

46 "About," Charm Industrial, accessed summer 2019, https://www.charmindustrial.com/about.

47 Davis et al., "Net-zero emissions energy systems."

48 See: Generation Atomic, "Carbon Neutral Fuels – Advanced Nuclear Reactors," YouTube video, May 31, 2018, https://www.youtube.com/watch?v=JFIbrqAkqnM; Bret Kugelmass, "Bret Kugelmass, at Colorado School of Mines, on Climate Change & Nuclear Energy," Titans of Nuclear, YouTube video, November 27, 2018, https://www.youtube.com/watch?v=3MfhZJqKgsU; and "Thesis," Energy Impact Center, accessed summer 2019, https://www.energyimpactcenter.org/thesis.

49 "Spot Prices," U.S. Energy Information Administration, accessed summer 2019, https://www.eia.gov/dnav/pet/pet_pri_spt_s1_a.htm.

50 Yoshitsugu Kojima, "Liquid Ammonia for Hydrogen Storage," *11th Annual NH3 Fuel Conference* (September 22, 2014), https://nh3fuelassociation.org/2014/09/06/liquid-ammonia-for-hydrogen-storage/.

51 Rong Lan and Shanwen Tao, "Ammonia as a suitable fuel for fuel cells," *Frontiers in Energy Research.* 2, no. 35 (August 28, 2014): https://doi.org/10.3389/fenrg.2014.00035.

52 About 5% if only minimal "fill in the gaps" case, perhaps substituting for jet fuel (1.5%) and a small portion of current

building heating fuel (total 6%) and agricultural energy use (1%); 40% if drop-in fuels became cheaper than fossil fuels and decarbonized transportation (15%), building fuel use (6%), industry fuel use (12%), agricultural energy use (1%), fossil fuel processing by displacing fossil fuels (4%), and escaped methane by displacing the extraction-side fugitive methane emissions (~2%).

53 "Measuring the Role of Deforestation in Global Warming," Union of Concerned Scientists, Updated December 9, 2013, https://www.ucsusa.org/global-warming/solutions/stop-deforestation/deforestation-global-warming-carbon-emissions.html.

54 "Mitigating Climate Change Through Coastal Ecosystem Management," The Blue Carbon Initiative, accessed summer 2019, https://www.thebluecarboninitiative.org/; and "Peatlands and Climate Change," International Union for Conservation of Nature, accessed summer 2019, https://www.iucn.org/resources/issues-briefs/peatlands-and-climate-change.

55 Paola Rosa-Aquino, "IPCC report: Planting trees isn't enough to save us from the climate crisis," *Grist*, August 9, 2019, https://grist.org/article/ipcc-report-planting-trees-isnt-enough-to-save-us-from-the-climate-crisis/; and Susan Minnemeyer, Nancy Harris, and Octavia Payne, "Conserving Forests Could Cut Carbon Emissions as Much as Getting Rid of Every Car on Earth," World Resources Institute, November 27, 2017, https://www.wri.org/blog/2017/11/conserving-forests-could-cut-carbon-emissions-much-getting-rid-every-car-earth.

56 Gabrielle Kissinger, Martin Herold, and Veronique De Sy, *Drivers of Deforestation and Forest Degradation: A Synthesis Report for REDD+ Policymakers* (Vancouver: Lexeme Consulting, 2012).

57 Minnemeyeret al., "Conserving Forests Could Cut Carbon Emissions."

58 See: "Amazon deforestation at highest level in 10 years, says Brazil," *Mongabay*, November 24, 2018, https://news.mongabay.com/2018/11/amazon-deforestation-at-highest-level-in-10-years-says-brazil/; and Letícia Casado and Ernesto Londoño, "Under Brazil's Far-Right Leader, Amazon Protections Slashed and Forests Fall," *New York Times*, July 28, 2019, https://www.nytimes.com/2019/07/28/world/americas/brazil-deforestation-amazon-bolsonaro.html.

59 See: Samuel McGlennon, "To save forests, keep an eye on agriculture," *Forests News*, March 4, 2016, https://forestsnews.cifor.org/40349/to-save-forests-look-at-farming-practices?fnl=en; Jennifer A. Burney, Steven J. Davis, and David B. Lobell, "Greenhouse gas mitigation by agricultural intensification," *Proceedings of the National Academy of Sciences* 107, no. 26 (June 29, 2010): 12052–12057, https://doi.org/10.1073/pnas.0914216107; and Dan Charles, "To Save the Planet, Give Cows Better Pasture," National Public Radio, February 13, 2017, https://www.npr.org/sections/thesalt/2017/02/13/514070632/to-save-the-planet-give-cows-better-pasture.

60 Minnemeyer et al., "Conserving Forests Could Cut Carbon Emissions."

61 Richard Waite, Tim Searchinger, and Janet Ranganathan, "6 Pressing Questions About Beef and Climate Change, Answered," World Resources Institute, April 8, 2019, https://www.wri.org/blog/2019/04/6-pressing-questions-about-beef-and-climate-change-answered.

62 "Why is enteric methane important?," Food and Agriculture Organization of the United Nations, accessed summer 2019, http://www.fao.org/in-action/enteric-methane/background/why-is-enteric-methane-important/en/; and "Understanding Global Warming Potentials," U.S. Environmental Protection Agency, accessed summer 2019, https://www.epa.gov/ghgemissions/understanding-global-warming-potentials.

63 "Key facts and findings," Food and Agriculture Organization of the United Nations, accessed summer 2019, http://www.fao.org/news/story/en/item/197623/icode/.

64 "Surf N'Turf"; and "Agriculture's Impact."

65 Neville Millar, "Management of Nitrogen Fertilizer to Reduce Nitrous Oxide Emissions from Field Crops," *Michigan State University Extension* E3152 (October 19, 2015): https://www.canr.msu.edu/resources/management_of_nitrogen_fertilizer_to_reduce_nitrous_oxide_emissions_from_fi; and "Understanding Global Warming Potentials.

66 "Global Greenhouse Gas Emissions Data," U.S. Environmental Protection Agency, accessed summer 2019, https://www.epa.gov/ghgemissions/global-greenhouse-gas-emissions-data.

67 Katy Dynarski, "Preventing Over-Fertilization for Better Crop Quality and Yield," *Teralytic* (blog), December 19, 2018, https://blog.teralytic.com/preventing-over-fertilization/; and "Corrective Measures and Management of Over-Fertilized Soils," The Center for Agriculture, Food, and the Environment, University of Massachusetts Amherst, updated March 16, 2017, https://ag.umass.edu/soil-plant-nutrient-testing-laboratory/fact-sheets/corrective-measures-management-of-over-fertilized.

68 Cornelius Oertel et al., "Greenhouse gas emissions from soils—A review," *Geochemistry* 76, no. 3 (October 2016): 327-352, https://doi.org/10.1016/j.chemer.2016.04.002.

69 Union of Concerned Scientists, *Agricultural Practices and Carbon Sequestration* (Cambridge: UCS, 2009).

70 Stephen Russell, "Everything You Need to Know About Agricultural Emissions," World Resources Institute, May 29, 2014, https://www.wri.org/blog/2014/05/everything-you-need-know-about-agricultural-emissions.

71 Gevan Behnke et al., "Long-term crop rotation and tillage effects on soil greenhouse gas emissions and crop production in Illinois, USA," *Agriculture, Ecosystems & Environment* 261 (July 1, 2018): 62–70, https://doi.org/10.1016/j.agee.2018.03.007.

72 Dan Charles, "To Save the Planet, Give Cows Better Pasture"; and Waite et al., "6 Pressing Questions About Beef and Climate Change, Answered."

73 Davis et al., "Net-zero emissions energy systems."

74 See: Davis et al., "Net-zero emissions energy systems"; Denise Brehm, "Nanoengineered concrete could cut carbon dioxide emissions," *MIT News*, January 30, 2007, http://news.mit.edu/2007/concrete; Kevin Bullis, "New Cement-Making Method Could Slash Carbon Emissions," *MIT Technology Review*, May 11, 2012, https://www.technologyreview.com/s/427906/new-cement-making-method-could-slash-carbon-emissions/; Kristin Majcher, "What Happened to Green Concrete?," *MIT Technology Review*, March 19, 2015, https://www.technologyreview.com/s/535646/what-happened-to-green-concrete/; and "Solidia," Solidia Technologies, accessed summer 2019, https://solidiatech.com/.

75 Davis et al., "Net-zero emissions energy systems"; Frédéric Simon, "Swedish steel boss: 'Our pilot plant will only emit water vapour'," *Euractiv,* May 11, 2018, https://www.euractiv.com/section/energy/interview/hybrit-ceo-our-pilot-steel-plant-will-only-emit-water-vapour/; and "Metal Oxide Electrolysis," Boston Metal, accessed summer 2019, https://www.bostonmetal.com/moe-technology/.

76 "Electric Arc Furnace," Industrial Efficiency Technology Database, accessed summer 2019, http://ietd.iipnetwork.org/content/electric-arc-furnace; and Davis et al., "Net-zero emissions energy systems."

77 Jim Robbins, "As Mass Timber Takes Off, How Green Is This New Building Material?," *Yale Environment 360,* April 9, 2019, https://e360.yale.edu/features/as-mass-timber-takes-off-how-green-is-this-new-building-material.

78 Robert F. Service, "Ammonia—a renewable fuel made from sun, air, and water—could power the globe without carbon," *Science Magazine,* July 12, 2018, https://www.sciencemag.org/news/2018/07/ammonia-renewable-fuel-made-sun-air-and-water-could-power-globe-without-carbon.

79 Rob Verger, "Aluminum production could get much better for the environment," *Popular Science,* May 10, 2018, https://www.popsci.com/aluminum-apple/.

80 J. L. Campos et al., "Greenhouse Gases Emissions from Wastewater Treatment Plants: Minimization, Treatment, and Prevention," *Journal of Chemistry* 2016, Article ID 3796352 (2016): https://doi.org/10.1155/2016/3796352; and "Basic Information About Landfill Gas," U.S. Environmental Protection Agency,

accessed summer 2019, https://www.epa.gov/lmop/basic-information-about-landfill-gas.

81 "Global Methane Budget: Highlights," Global Carbon Project, accessed summer 2019, https://www.globalcarbonproject.org/methanebudget/16/hl-compact.htm.

82 These are rough estimates based on the information presented on each topic in this book.

83 Estimated ranges based on analysis and sources presented in this book. Lower end corresponds with the pie chart in Chapter 3 showing remaining slices of emissions.

84 See: Sabine Fuss et al., "Negative emissions—Part 2: Costs, potentials and side effects," *Environmental Research Letters* 13, no. 6 (May 22, 2018): 063002, https://doi.org/10.1088/1748-9326/aabf9f; Martin et al., *Carbon Dioxide Removal Options*; and Bronson Griscom et al., "Natural climate solutions," *Proceedings of the National Academy of Sciences* 114, no. 44 (October 31, 2017): 11645–11650, https://doi.org/10.1073/pnas.1710465114.

85 "GDP Ranked by Country 2019," World Population Review, accessed summer 2019, http://worldpopulationreview.com/countries/countries-by-gdp/.

86 Griscom et al., "Natural climate solutions."

87 Paola Rosa-Aquino, "IPCC report"; and Minnemeyer et al. "Conserving Forests Could Cut Carbon Emissions."

88 Nikolas Hagemann et al., "Organic coating on biochar explains its nutrient retention and stimulation of soil fertility," *Nature Communications* 8, Article number 1089 (October 2017): https://doi.org/10.1038/s41467-017-01123-0; and Gevan Behnke et al., "Long-term crop rotation and tillage effects."

89 See uncertainties in Griscom et al., "Natural climate solutions."

90 Derived mainly from Griscom et al., "Natural climate solutions." Though higher numbers (2–6 Gt/yr for biochar and soil combined) are in Fuss et al., "Negative emissions—Part 2."

91 Martin et al., *Carbon Dioxide Removal Options*; and Oliver Milman, "Scientists say halting deforestation 'just as urgent' as reducing emissions," *The Guardian*, October 4, 2018, https://www.theguardian.com/environment/2018/oct/04/climate-change-deforestation-global-warming-report.

92 Intergovernmental Panel on Climate Change, *Revised 1996 IPCC Guidelines for National Greenhouse Gas Inventories: Reference Manual* (Geneva: IPCC, 1996).

93 Fuss et al., "Negative emissions—Part 2"; and Martin et al., *Carbon Dioxide Removal Options*.

94 Mahdi Fasihi, Olga Efimova, and Christian Breyer, "Techno-economic assessment of CO_2 direct air capture plants," *Journal of Cleaner Production* 224 (July 1, 2019): 957–980, https://doi.org/10.1016/j.jclepro.2019.03.086; and Richard Schiffman, "Why CO_2 'Air Capture' Could Be Key to Slowing Global Warming," *Yale Environment 360*, May 23, 2016, https://e360.yale.edu/features/pulling_co2_from_atmosphere_climate_change_lackner.

95 "Energy Transformation & Storage Alternatives," U.S. Naval Research Laboratory, accessed summer 2019, https://www.nrl.navy.mil/mstd/branches/6300.2/alternative-fuels.

96 "Climeworks," Climeworks, accessed summer 2019, https://www.climeworks.com/; "Direct Air Capture," Carbon Engineering, accessed summer 2019, https://carbonengineering.com/about-

dac/; and Fasihi et al., "Techno-economic assessment of CO_2 direct air capture plants."

97 Cristophe McGlade, "Commentary: Can CO_2-EOR really provide carbon-negative oil?," International Energy Agency, April 11, 2019, https://www.iea.org/newsroom/news/2019/april/can-co2-eor-really-provide-carbon-negative-oil.html; and Deepika Nagabhushan, "Leveraging Enhanced Oil Recovery for Large-Scale Saline Storage of CO_2," Clean Air Task Force, June 24, 2019, https://www.catf.us/2019/06/leveraging-enhanced-oil-recovery-for-large-scale-saline-storage-of-co2/.

98 Blake Eskew, "US Petrochemicals: The growing importance of export markets" (presentation, EIA Energy Conference, IHS Markit, June 4, 2018); and Phil De Luna (CERT), phone call with author, October 3, 2018.

99 Fuss et al., "Negative emissions—Part 2"; and David W. Keith et al., "A Process for Capturing CO_2 from the Atmosphere," *Joule* 2, no. 8 (August 15, 2018): 1–22, https://doi.org/10.1016/j.joule.2018.05.006.

100 Phil Renforth and Gideon Henderson, "Assessing ocean alkalinity for carbon sequestration," *Review of Geophysics* 55, no. 3 (September 2017): 636–674, https://doi.org/10.1002/2016RG000533; and Martin et al., *Carbon Dioxide Removal Options.*

101 Renforth and Henderson, "Assessing ocean alkalinity for carbon sequestration."

102 Renforth and Henderson, "Assessing ocean alkalinity for carbon sequestration."

103 Derek Martin et al., *Carbon Dioxide Removal Options* (Ann Arbor: University of Michigan, 2017); and Phil Renforth, e-mail message to author, July 15, 2019.

104 Martin et al., *Carbon Dioxide Removal Options.*

105 Holly Jean Buck, "The desperate race to cool the ocean before it's too late," *MIT Technology Review,* April 23, 2019, https://www.technologyreview.com/s/613327/the-desperate-race-to-cool-the-ocean-before-its-too-late/; and Martin et al., *Carbon Dioxide Removal Options.*

106 Buck, "The desperate race to cool the ocean"; Martin et al., *Carbon Dioxide Removal Options*; and Fuss et al., "Negative emissions—Part 2."

107 Keith et al., "A Process for Capturing CO_2 from the Atmosphere"; "Climeworks," Climeworks, accessed summer 2019, https://www.climeworks.com/; "Direct Air Capture," Carbon Engineering, accessed summer 2019, https://carbonengineering.com/about-dac/; and "Global Thermostat: A carbon negative solution," Global Thermostat, accessed summer 2019, https://globalthermostat.com/.

108 Larsen et al., *Capturing Leadership.*

PART THREE: HOW TO GET TO WORK

1 For example: "What is a Green New Deal?," Sierra Club, accessed summer 2019, https://www.sierraclub.org/trade/what-green-new-deal; and "Solutionary Rail - A people-powered campaign to electrify America's railroads and open corridors to a clean energy future," Solutionary Rail, accessed summer 2019, https://www.solutionaryrail.org/.

2 Such proposals have become essential components of US Democratic climate plans; for example, see: Maggie Astor, "Environmental Justice Was a Climate Forum Theme. Here's Why.," *New York Times*, September 5, 2019, https://www.nytimes.com/2019/09/05/us/politics/environmental-justice-climate-town-hall.amp.html; and Christine Ferretti, "Inslee details new 'Community Climate Justice' plan in Detroit," *The Detroit News*, July 29, 2019, https://www.detroitnews.com/story/news/politics/2019/07/29/inslee-visit-detroit-community-climate-justice-plan/1843084001/.

3 Masson-Delmotte et al., *Summary for Policymakers*.

4 Oliver Morton, *The Planet Remade: How Geoengineering Could Change the World* (Princeton: Princeton University Press, 2015).

5 In the graph, for temperature increase as of 2015, see: "Global warming reaches 1°C above preindustrial, warmest in more than 11,000 years," Climate Analytics, accessed summer 2019, https://climateanalytics.org/briefings/global-warming-reaches-1c-above-preindustrial-warmest-in-more-than-11000-years/.

6 David Keith, *A Case for Climate Engineering* (Cambridge: MIT Press, 2013).

7 Keith, *A Case for Climate Engineering.*

8 Keith, *A Case for Climate Engineering*; and Buck, "The desperate race to cool the ocean."

9 Buck, "The desperate race to cool the ocean."

10 Alexander A. Robel, Hélène Seroussi, and Gerard H. Roe, "Marine ice sheet instability amplifies and skews uncertainty in projections of future sea-level rise," *Proceedings of the National Academy of Sciences* 116, no. 30 (July 23, 2019): 14887–14892, https://doi.org/10.1073/pnas.1904822116; and Bethany Davies, "If all the ice in Antarctica were to melt, how much would global sea level rise? How quickly is this likely to happen?," *AntarcticGlaciers.org* (blog), accessed summer 2019, http://www.antarcticglaciers.org/question/ice-antarctica-melt-much-global-sea-level-rise-quickly-likely-happen/.

11 Buck, "The desperate race to cool the ocean."

12 See Simon Evans, "Direct CO_2 capture machines could use 'a quarter of global energy' in 2100," CarbonBrief, July 22, 2019, https://www.carbonbrief.org/direct-co2-capture-machines-could-use-quarter-global-energy-in-2100. I did my own estimates of emissions and sequestration pathways that achieved <350 ppm CO_2 by around 2080, and annual sequestration between 2060 and 2080 came close to 50 Gt/yr.

13 See: Jacqueline Ronson, "Will Genetically Modified Plants Save Us?," *Inverse,* January 11, 2017, https://www.inverse.com/article/26296-bunzl-genetically-modified-plants-geoengineering-climate-change; and Kim Krieger, "Genetically Engineered Bacteria Could Help Fight Climate Change," *Science Magazine,*

February 26, 2012, https://www.sciencemag.org/news/2012/02/genetically-engineered-bacteria-could-help-fight-climate-change.

14 Seth Motel, "Polls show most Americans believe in climate change, but give it low priority," Pew Research Center, September 23, 2014, https://www.pewresearch.org/fact-tank/2014/09/23/most-americans-believe-in-climate-change-but-give-it-low-priority/.

15 For example: Frédéric Simon, "Climate change will be key issue in EU elections, poll shows," *Euractiv,* April 16, 2019, https://www.euractiv.com/section/climate-environment/news/climate-will-be-key-issue-in-eu-elections-poll-shows/; Holly Burke, "New Poll: Primary Voters Vault Climate to Top Tier Issue," League of Conservation Voters, February 14, 2019, https://www.lcv.org/article/new-poll-primary-voters-vault-climate-top-tier-issue/; and SSRS poll for CNN, Tuesday, April 30, 2019, http://cdn.cnn.com/cnn/2019/images/04/29/rel6a.-.2020.democrats.pdf.

16 Andrew Chatzky and James McBride, "China's Massive Belt and Road Initiative," Council on Foreign Relations, updated May 21, 2019, https://www.cfr.org/backgrounder/chinas-massive-belt-and-road-initiative.

17 Eduardo Porter, "Does a Carbon Tax Work? Ask British Columbia," *New York Times,* March 1, 2016, https://www.nytimes.com/2016/03/02/business/does-a-carbon-tax-work-ask-british-columbia.html; and Kathryn Harrison, "Lessons from British Columbia's carbon tax," *Policy Options,* July 11, 2019, https://policyoptions.irpp.org/magazines/july-2019/lessons-from-british-columbias-carbon-tax/.

18 See "CSIRO," CSIRO, accessed summer 2019, https://www.csiro.au/; Robert F. Service, "Ammonia—a renewable fuel made from

sun, air, and water—could power the globe without carbon," *Science* (blog), July 12, 2018, https://www.sciencemag.org/news/2018/07/ammonia-renewable-fuel-made-sun-air-and-water-could-power-globe-without-carbon; Trevor Brown, "Renewable ammonia demonstration plant announced in South Australia," Ammonia Industry, February 16, 2018, https://ammoniaindustry.com/renewable-ammonia-demonstration-plant-announced-in-south-australia/; and Lexy Hamilton-Smith, "Hydrogen fuel breakthrough in Queensland could fire up massive new export market," *Australian Broadcasting Commission,* August 7, 2018, https://www.abc.net.au/news/2018-08-08/hydrogen-fuel-breakthrough-csiro-game-changer-export-potential/10082514.

19 Zoya Terstein, "The group that pushed the Green New Deal sets its sights on 2020 and beyond," *Grist,* June 19, 2019, https://grist.org/article/for-the-sunrise-movement-2020-is-just-the-beginning/.

20 See: "Citizens' Climate Lobby," CCL, accessed summer 2019, citizensclimatelobby.org.

21 Robert Walton, "Google: Powering data centers with clean energy 24/7 'no easy feat'," *Utility Dive,* October 12, 2018, https://www.utilitydive.com/news/powering-data-centers-with-clean-energy-247-no-easy-feat-google-finds/539497/.

22 H. K. Gibbs et al., "Brazil's Soy Moratorium," *Science* 347, no. 6220 (January 23, 2015): 377–378, https://doi.org/10.1126/science.aaa0181.

23 Based on conversations with solar and nuclear innovators, a solar or battery chemistry could be proven for around $200,000, while a nuclear design would require hundreds of millions of dollars.

24 Based on 2010 cost of Chinese coal and methane plants from Danwei Zhang and Sergey Paltsev, *The Future of Natural Gas in China: Effects of Pricing Reform and Climate Policy*, (MIT Libraries, March 2016), https://dspace.mit.edu/handle/1721.1/103778; and converting capital costs per unit of peak capacity to capital costs per unit of annual output (the more relevant metric for driving deployment decisions) using capacity factors from: "Electric generator capacity factors vary widely across the world," U.S. Energy Information Administration, September 8, 2015, https://www.eia.gov/todayinenergy/detail.php?id=22832 shows that nuclear plants around $900/kW would be cheaper than or about the same cost as the cheapest coal and methane plants in the world in terms of cost to add an extra amount of annual generation output.

25 Davis et al., "Net-zero emissions energy systems."

26 Michael Sivak and Brandon Schoettle, "Relative Costs of Driving Electric and Gasoline Vehicles in the Individual U.S. States," University of Michigan, January 2018, www.umich.edu/~umtriswt/PDF/SWT-2018-1_Abstract_English.pdf.

27 See: "Joint Center for Energy Storage Research," JCESR, accessed summer 2019, http://www.jcesr.org/; and "New York Battery and Energy Storage Technology Consortium," NY-BEST, accessed summer 2019, https://www.ny-best.org/.

28 See: Bertel Schmitt, "Exclusive: Toyota Hydrogen Boss Explains How Fuel Cells Can Achieve Corolla Costs," *The Drive*, January 18, 2019, https://www.thedrive.com/tech/26050/exclusive-toyota-hydrogen-boss-explains-how-fuel-cells-can-achieve-corolla-costs.

29 "Air conditioning use emerges as one of the key drivers of global electricity-demand growth."

30 Davis et al., "Net-zero emissions energy systems"; Clean Air Task Force, *Decarbonizing Cement and Steel: Key understandings and research and action needed* (2019); Niall McCarthy, "China Produces More Cement Than the Rest of the World Combined [Infographic]," *Forbes*, July 6, 2018, https://www.forbes.com/sites/niallmccarthy/2018/07/06/china-produces-more-cement-than-the-rest-of-the-world-combined-infographic/#717ade1f688; and Eustance Huang, "Fortescue's CEO is seeing 'strong growth' in China's steel production," *CNBC*, May 1, 2019, https://www.cnbc.com/2019/05/01/fortescue-ceo-sees-strong-growth-in-chinas-steel-production.html.

31 "The cost of carbon capture: is it worth incorporating into the energy mix?," *Power Technology*, October 4, 2018, https://www.power-technology.com/features/carbon-capture-cost/.

32 Martin et al., *Carbon Dioxide Removal Options*.

33 "Carbon Fee and Dividend Policy and FAQs," CCL, accessed summer 2019, https://citizensclimatelobby.org/carbon-fee-and-dividend/.